A Green and Global Europe

A Green and Global Europe

Nathalie Tocci

polity

First published in 2023 by Polity Press

Polity Press
65 Bridge Street
Cambridge CB2 1UR, UK

Polity Press
111 River Street
Hoboken, NJ 07030, USA

ISBN-13: 978-1-5095-5516-1
ISBN-13: 978-1-5095-5517-8 (pb)

A catalogue record for this book is available from the British Library.

Library of Congress Control Number: 2022937771

Typeset in 10.5 on 12.5pt Sabon
by Fakenham Prepress Solutions, Fakenham, Norfolk NR21 8NL
Printed and bound in the UK by CPI Group (UK) Ltd, Croydon

The publisher has used its best endeavours to ensure that the URLs for external websites referred to in this book are correct and active at the time of going to press. However, the publisher has no responsibility for the websites and can make no guarantee that a site will remain live or that the content is or will remain appropriate.

Every effort has been made to trace all copyright holders, but if any have been overlooked the publisher will be pleased to include any necessary credits in any subsequent reprint or edition.

For further information on Polity, visit our website:
politybooks.com

Contents

Abbreviations

bcm	billion cubic metres (of gas)
BRI	Belt and Road Initiative
CAI	Comprehensive Agreement on Investment between the EU and China
CBAM	Carbon Border Adjustment Mechanism
CCS	carbon capture and storage
COP	Conference of the Parties
DG	Directorate General of the European Commission
DRC	Democratic Republic of Congo
EEAS	European External Action Service
EIB	European Investment Bank
ETS	Emissions Trading Scheme
EU	European Union
FTAs	free trade agreements
GDP	gross domestic product
HRVP	High Representative of the Union for Foreign Affairs and Security Policy and Vice President of the European Commission
IEA	International Energy Agency
IMF	International Monetary Fund
IPCC	Intergovernmental Panel on Climate Change
IRENA	International Renewable Energy Agency
LNG	liquefied natural gas
NATO	North Atlantic Treaty Organization
NDCs	nationally determined contributions

NDICI	Neighbourhood, Development and International Cooperation Instrument
NIMBY	not in my back yard
OECD	Organization for Economic Cooperation and Development
OPEC	Organization of the Petroleum Exporting Countries
PV	photovoltaic
TAP	Trans Adriatic Pipeline
UNDP	United Nations Development Programme
UNFCCC	United Nations Framework Convention on Climate Change
WTO	World Trade Organization

Preface

I began working on Europe in the heyday of the integration project, at the height of the liberal international order. The euro was about to be launched, and the Union was on the cusp of its 'big bang' enlargement that would reunify the continent after decades of Cold War divide. Europe was associated with values, opportunity and even power. Whereas it still, at times unknowingly, relied on the American security umbrella, the EU genuinely looked like a different kind of power, a superpower in its own right. It was the physical and political space to which many aspired, and which I felt so fortunate to live and work in.

Beginning with the global financial crisis, that picture changed dramatically. Europe became a land of division, inequality and restriction; the dream of European integration started becoming the nightmare of many. The Union held together, but languished, failing to find a way out of a protracted crisis. Those were years in which I focused all my energies on Europe's global role, convinced as I am that in an increasingly contested, connected and complex world, Europeans represent a community of fate. We often disagree with one another, and at times appear not to share either interests or values. But in the twenty-first century, we have a peaceful and prosperous future only if we stick together in the wider world. Russia's brutal invasion of Ukraine, harking back to the darkest days of the twentieth century, made this clearer than ever.

Against the backdrop of the pandemic and war in Europe, with the renewed sense of solidarity amongst Europeans these brought about, Europe's green agenda has given the Union a new lease of life. I firmly believe that a green Europe is the key to the EU's revival. Without it, all else risks losing meaning given both the strategic urgency of energy security and the existential nature of the climate crisis. Through its green agenda, Europe can rejuvenate strategically, economically and politically, if it gets it right. A green Europe represents a normative, strategic, economic and political project all rolled into one.

This said, I remain wholeheartedly convinced that Europe's rationale in the twenty-first century can only be global in nature. Nowhere is this clearer than on climate change and the energy transition, the topic of this book. Getting a green Europe right means becoming global too.

I have followed energy and climate for several years now, but this is the first book I have written about the subject. I have sat on the board of energy companies as an independent non-executive member since 2013, but until recently, I unconsciously fell into the same trap that I argue against in the conclusion of this book. My work on policy, research and on boards was kept artificially siloed and separate.

As I became energized about a green Europe and the promise it represents, while also observing that we remain far from walking the walk of a global Europe, I broke out of my own mental silos, and brought what I had learned in one area of my professional life into the others. The idea for this book was the result.

Translating that idea into *A Green and Global Europe* was possible, however, only with the help of many, whom I wish to thank. Heartfelt thanks to Nick Burns and Karl Kaiser for encouraging me to come to Harvard Kennedy School, and above all to Pierre Keller for making my visiting professorship possible. My time in Cambridge, MA, while far too short, gave me the mental space, and above all the inspiration to write this book. Thanks to Erika Manouselis for making me feel immediately at home, Paul-Étienne Pini for his impeccable research assistance, and to Diego

Garcia Blum and all my students for making my time so special. Thanks to Luca Franza, Margherita Bianchi, Marco Giuli, Pierpaolo Raimondi and Lorenzo Colantoni at the Istituto Affari Internazionali, to Heather Grabbe, Simone Tagliapietra and Alexandros Yannis for their invaluable comments and suggestions. Special thanks go to Claudio Descalzi and Eni's management for teaching and inspiring me so much. This book would not have been possible without all I learned from them.

I also wish to thank wholeheartedly Vivien Schmidt for opening my eyes to the fact that Polity would be the ideal home for my work, to Mary Kaldor for making the connection, and to Polity's anonymous reviewers for their insights and suggestions. Thanks to Louise Knight and Inès Boxman for being so supportive, responsive and a real pleasure to work with all along the way.

And finally and most importantly, my biggest thank you goes to my family, Kike and Diego, my anchor, love and joy. It is to them that this book is dedicated.

Rome, March 2022

Introduction

The European Union is exiting almost two decades of existential crisis. From the rejection of the Constitutional Treaty, to the sovereign debt crisis that risked tearing the Eurozone apart, the migration crisis that laid bare the lack of European solidarity, and the Brexit referendum that raised the spectre of a domino effect across a Eurosceptic Europe, the Union has been teetering on the brink for too long.

As it did, the EU progressively lost its narrative: it no longer had a compelling story to tell. The narrative of peace on the continent after the death and destruction of two world wars no longer resonated with younger generations, while that of European prosperity through the single market and currency rang hollow after the global financial crisis and its exposure and exacerbation of socio-economic disparities within and between Member States (Tocci 2019a; 2020). Rather than being associated with growth, solidarity and opportunity, 'Brussels' became synonymous with austerity, inequality and unaccountability in the eyes of many Europeans.

As the 2010s came to a close, Europe understood the need to ratchet up its climate agenda, presenting its new trademark: the European Green Deal. The louder alarm bells from the science community, the growing public pressure from Greta Thunberg's Fridays for Future movement, and the strong showing of the green parties at the 2019 European Parliament elections, all pointed in the same direction: just like the cover of this book, the European flag had to be painted green.

When Covid-19 hit, the fear was that the European Green Deal would be shelved, and the EU as a whole would founder. Had the EU failed to rise to the challenge posed by the pandemic, it would have been a crisis too many for the European project to endure. If 'Brussels' had reconfirmed its image as one of disunity, detachment and lack of empathy, the entire European edifice that cemented peace on the continent after centuries of violence could have tipped over the point of no return.

The EU did not fail. While slow to start, the Union navigated the crisis as one, be it in terms of public health, notably vaccination rollout, and economic recovery (Alcaro and Tocci 2021). Through its response, the EU turned the pandemic crisis into the proverbial opportunity, putting flesh onto the bones of its new trademark: the European Green Deal (European Commission 2019). Russia's invasion of Ukraine and the debate this reignited on energy security and Europe's energy independence propelled the strategic rationale of decarbonization to new heights. As it exits years of protracted crisis, a green Europe represents a normative vision to fight the existential crisis posed by anthropogenic climate change; an economic growth strategy that pursues decarbonization through innovation, job creation, industrial capacity and reduced inequalities; a strategic imperative to ensure energy security; and a route to a political Union by fostering a common cause between Member States and by reconnecting to the European public, especially youth. By painting its flag green, the EU has found the recipe to help save the planet while reviving itself politically.

Precisely because it is so existential for the future of Europe, getting both the story and the practice right is crucial. This hinges on a successful energy transition, given that energy accounts for 75 per cent of Europe's greenhouse gas emissions (European Commission 2021a). Europe's energy transition lies at the core of a green and sustainable Europe. It is also a key pillar of a green and sustainable world. With 800 million people still lacking access to electricity and 2.6 billion people to clean cooking solutions, and with the global population set to increase by 2 billion over the next three decades, the current energy system is both environmentally unsustainable and socio-economically insufficient. Only a more modern, decarbonized and accessible energy system can square the circle of sustaining people and planet alike. The European ambition is to lead the way in this epochal change.

A technology- and policy-driven transition

This is a tall order: previous energy transitions took place over decades, at times centuries, featuring the gradual replacement of one energy source by another, be it wood, coal, or oil and gas. The transition from wood to coal took place between 1790 and 1845; that from coal to oil happened between 1865 and 1930. All transitions were kick-started by technological breakthroughs and proceeded through slow and non-linear changes (IEA 2019), in which a new energy source was added to existing ones, rather than immediately replacing them.

On these long energy journeys, technological breakthroughs were always necessary but alone they were insufficient. Political and business decisions taken at specific times and places accelerated or slowed down the pace of change along the way. British Prime Minister Winston Churchill's decision to switch the main power source of the Royal Navy from coal to oil; the European decision to buy Soviet gas after the 1973 oil price shock; or the decision taken by US car manufacturers to veer away from electric vehicles in the early 1900s were three such moments.

Technological breakthroughs coupled with political or business choices then faced powerful sources of resistance against change, explaining the duration and erratic nature of past transitions. Scale, sunk investments, incumbents' vested interests, consolidated networks and supply chains, stubborn behavioural patterns, and an uneven playing field between potential losers who are typically more motivated to resist change than potential winners to promote it, all added up, decelerating change.

Furthermore, with economic development proceeding at different paces in different parts of the world, energy transitions have been remarkably differentiated. In the past, as electricity began displacing kerosene, the latter received an unexpected boost with the development of the aviation industry, which lives on to this day. More recently, while coal consumption started dropping in Europe, followed by the United States, the rise of China and emerging economies provided it with a new lease of life. The sheer abundance of cheap coal in energy-hungry economies explains why it is so hard to wean off the dirtiest fossil source of all. Other energy sources such as nuclear may also witness a revival in the years ahead. As energy demand rises, technology improves and the burning of hydrocarbons reduces, nuclear energy may well have a future in those countries where public opinion considers it a politically acceptable piece of the decarbonization puzzle. In fact, with the exception of the transition away from biomass to fossil fuels, all transitions have been gradual, differentiated and marked by ebbs and flows.

On one level, the current transition is no different, revolving around the replacement of fossil fuels with low-carbon energy sources in view of technological breakthroughs and political decisions. With the strides forward made in the research, development and production of green technologies, alongside their rapidly falling costs, we are past the stage at which the question is whether the transition to a decarbonized energy system will eventually happen. With the greater availability and declining costs of clean technologies, it will simply make increasing economic sense to decarbonize in the years ahead. This energy transition will happen, just like other transitions have taken place in the past. The question at stake today is not whether the transition will happen, but how it will unfold and how long it will last.

On another level, however, this transition is unprecedented. The magnitude of change is incomparably greater than in the past. Dwarfing the past, today's world economy is worth approximately $86 trillion. It is expected to reach around $185 trillion by the time we reach net zero greenhouse gas emissions by mid-century (Yergin 2021). While renewables are gradually making their way into the global energy mix, 80 per cent of all energy sources still come from hydrocarbons, not to mention the use of fossil fuels to produce plastics and a myriad other petrochemical derivatives. In less than three decades, we will need to fuel a more than doubled economy with a significantly larger population, with hydrocarbons dropping to around 20 per cent of the total energy mix if we are to reach net zero (IEA 2021a). Unlike the past, in which new energy sources were added onto old ones, the aim today is essentially to replace one energy system with another.

The implications are huge. The breadth and depth of the infrastructural, institutional, economic, cultural and behavioural changes that would come along with this transition are unlike anything we have seen before. The world today is demographically larger and incomparably more networked than in times of previous transitions. Power within it is more diffuse, and the number of players that must be brought into the picture far greater. The current energy transition will involve millions of public and private institutions and billions of people, who will need to change how they buy, work and live. It will require mindboggling sums of money, technical skills and organizational capacities. The complexity embedded in the current transition is unlike anything we have seen before.

Making the current transition unprecedented is not just the magnitude and complexity of change, but also the lightning speed at which it should happen. Driven not only by technological progress and one-off political choices but also by the solid scientific evidence on, and the growing public awareness of, the anthropogenic nature of climate change and its effects, the speed at which this transition *should* happen is incomparably higher than at any time in the past. Today we simply do not have the luxury to wait for many decades and experience a wildly heterogeneous transition that proceeds in fits and starts in different parts of the world. This transition has a deadline: it needs to take place over the next three decades to reach net zero emissions. This is nothing less than revolutionary.

This means that, unlike its predecessors, this energy transition will need to be politically steered. It cannot rely mainly on technology, time, markets and serendipity to do the job. One needs only to think that some technologies – for instance concerning hydrogen or carbon capture and storage (CCS) – have been known for many years. Left to their own devices, they would take decades to come online, and many would probably never see the light of day. What makes them not only necessary but also feasible today is the growing appreciation of climate change, the normative goal the world has given itself to reach net zero emissions, and the ensuing policies necessary to get there, such as carbon pricing or green finance taxonomies. Only a political and policy-driven energy transition can be fast enough to address the existential crisis posed by climate change.

A green and global Europe

Past energy transitions gave rise to profound economic, societal, political and ideational changes. Against the backdrop of the envisaged magnitude, complexity and speed of the current transition, one can reasonably expect that its social, economic, political and geopolitical consequences will be even more acute this time (Hafner and Tagliapietra 2020). Amidst these changes, there will be many winners, but there will be losers too. And losers will make their voices heard. In fact, resistance to this transition is likely to be more vocal, targeted and therefore more effective than in the past. This is precisely because of the political and policy-driven nature of this transition. It is one thing to oppose nameless technological changes or amorphous market forces. Quite another is resisting change that is perceived as being driven by specific institutions, political forces and economic interests. The potential losers of decarbonization within and between countries will speak out, weighing in on those decision-makers on whose policies the energy transition crucially depends. Hence, the need to make this energy transition socially, economically and therefore also politically and geopolitically sustainable is more important than ever.

This has profound implications for Europeans and for the European Union. The energy transition to decarbonize Europe will reverberate deeply across societies and economies, both within and beyond EU borders (Leonard et al. 2021). Having rightly elevated a green Europe into its new vision and strategy, the EU will be viewed as the political and policy home of this epochal change, both by European citizens and by third countries. It will be applauded by the potential winners of the energy transition and denounced by its losers, with the latter making their opposition loud and clear. This means that it is essential to make a green Europe socially, economically and therefore also politically and geopolitically acceptable. In fact, given the political nature of this transition, its political acceptability is a prerequisite for a green Europe to actually materialize, and thus for the EU to embed its new raison d'être in it. Preparing and acting both on the energy transition *and* on its social, economic and therefore political and geopolitical consequences is essential. Unless this is carried out, the EU's vision will either fail to translate into practice, falling short of its responsibility to people and planet alike, or inadvertently unleash backlashes across countries and policy areas that could come biting back at the Union.

Such potential backlashes are magnified by the fact that the transition will take place at different paces and in different ways across different parts of the world. One only needs to think of the vastly different levels of climate and energy awareness, capacities and governance models that exist today in Europe, the US, China, Russia, the Gulf or emerging economies in Africa and Asia. The EU's climate consciousness is matched only by that of some developing countries in Africa and Asia as well as small island states, which bear none of the responsibility that Europe does for greenhouse gas emissions, let alone historical ones. In the US, climate awareness is gradually rising, but climate action remains highly contentious politically. China, too, is increasingly cognizant of the need to act on climate but is unwilling to compromise its economic growth to serve this end. This is even truer of India and much of the developing world, which fiercely objects to the climate injustice of having to bear the cost, in terms of forsaken development, of a climate crisis they played virtually no role in creating. With hundreds of millions worldwide still having no access to energy at all, most of whom live in Africa, the energy debate in many developing countries bears little resemblance to that on the European Green Deal. It rather revolves around energy access and the need to increase natural gas use to save the millions of lives lost annually to indoor air pollution generated by charcoal and kerosene cooking stoves.

These inevitably differentiated energy transitions worldwide have two sets of implications for Europe that will be unpacked in this book. Internally, as the EU decarbonizes faster than other world regions, those Europeans who fear losing out from the energy transition will increasingly point to the slower – and in their view less disruptive – pace of change elsewhere. While Europeans shut down their coal plants, raise carbon prices hindering the competitiveness of their carbon-intensive industries, and debate whether nuclear energy and natural gas have a future in the continent's transition, the US resists pricing emissions, while China or India build coal plants, while ramping up gas use precisely to decarbonize their rapidly growing economies. Those segments of European society, from industry to individuals, that resist change because they fear being harmed by it, will use the differentiated energy transitions worldwide as ammunition for their cause. If the world as such must decarbonize in order to address the climate crisis, what good does it do either to Europe or to the world if the EU speeds towards net zero, with all the socio-economic disruptions this could cause, if others proceed at a far more leisurely pace?

Externally, as Europe decarbonizes faster than other regions, it runs the risk of priding itself on being a global climate leader, but one which has fewer and fewer followers. This would make no sense, either from a normative or from an interest-based perspective.

Viewed through a normative lens, a green Europe only makes sense if embedded in a decarbonizing world. Today, the EU is ahead of the global green curve. It represents under 8 per cent of global emissions and that figure is fast declining.[1] It is the only global player with a plan to reach climate neutrality, with a clear pathway for the crucial 2020s ahead of us. Whether that plan is the best possible one, or whether it will be implemented as foreseen, remains to be seen. However, what can be safely said is that the European Green Deal is the only plan in town. Both China and the US, the world's two largest emitters, at 28 and 14 per cent, respectively, have made bold pledges and announced major policy initiatives,[2] but still lack the implementing laws, regulations and funds to get there (Sims Gallagher 2021). China has committed to climate neutrality by 2060, established an embryonic carbon market, and announced an end to international coal plant projects. It also leads the global way on many renewable technologies, notably solar, and has made huge strides on electric vehicles. However, China lacks a climate law, the reform of its power sector is sluggish, and its domestic coal use proceeds apace and largely unabated. The US has recommitted to the Paris Agreement and pledged climate neutrality by 2050. President Joe Biden's administration is determined to put its money where its mouth is on the energy transition. However, the US administration's ambitions have been watered down by Congress, carbon pricing is far off the horizon, and many of the regressive steps taken by the Trump administration are not easily reversed. The EU is far ahead of the US and China, to name only the major players, in its energy transition policies. However, given the global public good nature of climate, the EU's ambition and action, if not followed by others, are no reason to rejoice. The European Green Deal represents a normative vision only if it succeeds both in making Europe the first net zero continent *and* spurs the rest of the world to follow suit.

Hard-nosed interests point to the same conclusion. Were the EU to decarbonize while failing to bring the rest of the world along with it, its global competitiveness would suffer, its industrial base would be hollowed out, and the potentially regressive socio-economic effects of the transition would be exacerbated. Whereas a decarbonized economy in future may be more prosperous, a decarbonizing one amidst a fossil-fuelled world would likely not. The EU would risk inadvertently contributing to the decoupling of global supply chains between green and brown economies, with higher costs for all. The EU will successfully decouple emissions and prosperity in Europe only if the rest of the world follows a similar path. The nature and pace of transitions across different world regions will differ, but the direction of travel must be the same.

Whether viewed through the lens of internal European social, economic and therefore political acceptability, or through the lens of external norm diffusion and interest-based pursuit, a green Europe makes sense only if it is a global one too. Concomitantly sustaining the transition internally, incentivizing change beyond borders, and remaining the normative anchor of international climate action will not be easy. It is only possible if a green and global Europe are seen as two sides of the same coin.

Structure of this book

The chapters that follow develop this argument in three successive steps. The first step focuses on the internal European level, explaining the trajectory of Europe's energy and climate policies and revealing the social, economic and political prerequisites for their sustainability and success.

Chapter 1 lays out the gradually converging policy and political paths of climate and energy in the European Union. Beginning in the 1990s, the EU began affirming itself as a global climate leader. The need to act on climate change was both unanimously shared by Member States, especially before the enlargement to central and eastern Europe, and largely off the radar of their domestic political debates. The high level of policy consensus and low political salience of the subject enabled the European Commission to affirm itself as a key player in international climate talks. Energy policy followed a different trajectory. Member States have traditionally had different energy interests, goals and perspectives. Only with the Treaty of Lisbon did supranational

European institutions acquire residual competences, largely aimed at promoting energy efficiency, decarbonization and, above all, energy security. Divergent strategic and economic interests continued placing structural limits on European energy integration, although steps forward were made under the political radar to increase the EU's overall energy security, notably through regulation and infrastructure development. Moreover, the bifurcated trajectories of climate and energy policy contributed to explaining why the EU's own emissions reduction targets were not met.

The Energy Union started bringing the climate and energy tracks more explicitly within the same overarching policy framework. However, it is only more recently, with the European Green Deal, that climate and energy have moved front and centre of European political life and the EU has made its first steps towards integrating climate across different institutions and policy areas. This is a prerequisite to meeting the EU's own climate targets in practice, with energy policy becoming increasingly driven by climate policy rather than proceeding in parallel with it. With the European Green Deal, the foundations of an all-of-government approach are being laid.

After briefly unpacking the European Green Deal, including its legislative, regulatory and financial components, this chapter discusses how climate and energy are rising on the domestic political agenda across different European countries. The war in Ukraine and the realization that Europeans should urgently wean themselves off Russian fossil fuels increased further the political salience of the Green Deal, highlighting the strategic imperative to reconcile energy security and decarbonization through a strategy that blends principle and pragmatism, with a carefully calibrated balance that changes over time. All this suggests that the EU will become increasingly politicized in the years ahead. The chapter concludes by making the distinction between a politicized and a political EU, explaining how a green Europe holds the promise of a political and thus more integrated, united and legitimate Europe, but also why navigating this journey will be highly complex.

Chapter 2 opens by discussing the crisis of liberal democracy in the West, particularly the double attack it has faced in recent years both by authoritarian countries and by nationalist populists, which in Europe have been invariably Eurosceptic. It highlights the potentially adverse distributional effects of the energy transition, and the way these are providing new ammunition for Eurosceptic forces. The chapter then zooms in on the 2021–2 energy crisis as a canary in the transitional coalmine, explaining why this crisis, and the volatility underpinning it, may indicate the future trends ahead. It concludes by discussing how the potentially regressive effects of the transition could be addressed, including through funding mechanisms, joint European strategic reserves and greening long-term relationships with producer countries, as well as why strategic communication should focus on countering disinformation about the transition and engage in positive storytelling.

The second step in this book's analysis broadens the analysis to the regional level, with chapter 3 unpacking the multiple ways in which the EU's climate and energy policies will reverberate across its troubled neighbourhood. Europe's surrounding regions, both to the east and south, are rife with conflicts, crises and multifaceted fragilities. The EU's energy transition will impact these in different ways, in some cases generating new risks and challenges to be addressed, and in others unlocking green development opportunities to be reaped.

For one, Europe's historical greenhouse gas emissions have played no small role in the current climate crisis, which increasingly interlocks with existing vulnerabilities in Europe's surrounding regions like the Sahel and Iraq. Both are amongst the world's regions most affected by climate change, with global average temperatures expected to rise by 5°C. Both suffer from deep political, social, economic and security vulnerabilities, which the climate crisis is exacerbating greatly. Yet to date, the EU's attention to climate adaptation has been marginal in the broader articulation of its foreign policies towards these regions, highlighting the need to double down on climate adaptation, putting this at the core rather than the margins of European foreign policy.

Another crucial element revolves around countries that produce fossil fuels, which may be amongst the losers of the transition. Many of these are located in Europe's surrounding regions, and some are highly dependent on their fossil fuel exports to Europe. As the EU gradually transitions away from oil and gas, it may indirectly exacerbate the vulnerabilities of these countries. Under the surface, however, the story is nuanced. Whereas some fossil fuel-rich states have the necessary financial buffers and more time to diversify and mitigate, others are not only more fragile but also more likely to be hit first by Europe's transition. Countries like Algeria, Libya and Nigeria will require particular attention. As the EU transitions away from fossil fuels, supporting these countries in their own climate mitigation policies will be essential. This requires both public funding and, more importantly given the magnitude of investment needed, creating the institutional and financial rules and incentives for private players to invest in mitigation projects in these countries. Other fossil fuel states have instead weaponized their natural resources to reap strategic gain. Nowhere is this clearer than in the case of Russia, a country with regard to which the EU's energy security and its decarbonization agenda, including through the diversification and greening of energy relationships, are intertwined.

Just as important are the vulnerabilities that may arise not as an indirect effect of Europe's energy transition but as the direct consequence of the EU's climate policies. In particular, the Carbon Border Adjustment Mechanism (CBAM), designed to avoid carbon leakage as the EU develops and expands its Emissions Trading Scheme (ETS), will affect critically countries like Turkey with whom the EU already has problematic relations, as well as developing countries that may lack the industrial, technological and governance capacities to dodge the levy. In the case of Turkey, the EU–Turkey customs union poses an additional complexity. As CBAM comes into effect, the continued functioning of the customs union can only be assured if Turkey moves towards carbon pricing.

At the same time, Europe's energy transition also opens the prospect of green development opportunities beyond its borders. This is particularly true for two sets of fossil fuel-poor countries: those like Morocco, which could become green champions, and vulnerable fossil fuel transit countries like Ukraine, whose energy transition could support the country's weaning itself off gas transit fees from Russia and strengthen its overall resilience as a result. The onus is on the EU to support the flourishing of these green opportunities, strengthening sustainable development and resilience beyond its borders.

The third step in this book's analysis turns to the global level. Chapter 4 begins by discussing how two key geopolitical trends shaping the twenty-first century – the US–China rivalry and the question of globalization – interlock with the energy transition, feeding into these trends. In a decarbonized world, energy power is likely to move down the value chain, with a premium on technology over raw materials. The latter will still matter – notably critical minerals – but technological and economic strength, as well as governance systems, will play a much bigger role. This suggests a growing convergence between geopolitical power and power in the energy sphere, with energy consequently moving to the core of great power politics. A green world is also likely to be more decentralized and regionalized. This will both 'democratize' energy, but also reinforce trends towards reshoring, nearshoring, as well as closure, decoupling, protectionism and deglobalization. All this plays into three sets of global cleavages that will fundamentally shape the energy transition and multilateral climate diplomacy: the West–China rivalry, the transatlantic partnership and relations between the Global North and South.

The EU has set out the ambition to become strategically autonomous, particularly as regards the economy and industry, including energy. The energy transition certainly holds the promise to strengthen European autonomy and resilience through the development of green industrial ecosystems. However, it also generates new vulnerabilities, particularly in relation to China, as well as the complex trade-offs the EU will have to make, including between human rights, sustainable development and its net zero goal.

While there is no magical way to square the circle between the EU's climate neutrality, autonomy, prosperity and values, the best available recipe includes the development of green partnerships with different world regions, beginning with the US. Under the Biden administration, this appeared within reach, but scratching the surface the gap separating Europe and the US remains wide. There are significant differences over industrial policy, carbon pricing and green taxonomies governing sustainable investments, which the EU and the US should bridge, with the EU demonstrating greater flexibility over carbon pricing and the US greater openness towards green taxonomies. Both carbon pricing and taxonomy are crucial to generate the institutional, regulatory and financial incentives for the corporate world to accelerate along the transition path. However, the effectiveness of these policies hinges on the extent to which their reach broadens beyond the EU and across the Atlantic. If Europe and the US move together on these issues, they will likely create a critical mass that others – notably China – will find difficult to resist.

A final cleavage arising from Europe's transition is that between the Global North and South, which is linked to the crucial question of climate justice. While bedevilling international climate negotiations since the onset of the Conference of the Parties (COP), the Global North and South divide could widen even further as the energy transition moves from rhetoric to reality. The technical complexity and financial prerequisites underpinning climate action require the mobilization of incredible sums in the years ahead. The EU and European countries, including the UK, collectively represent the lion's share of global public finance, but this still falls short of the $100 billion per year set out in the 2015 Paris Agreement, not to mention trillions needed to reach net zero and the even higher sums if 'loss and damage' is taken seriously. Furthermore, beyond the gross sum, there are tough questions that Europeans need to address concerning how and on what such funds should be spent. Currently the bulk of climate finance is invested in the West. However, the energy transition will require trillions of extra funds, most of which should be spent in developing and emerging economies. Critical in this respect is the mobilization of private funds, which raises the question of how policy can ensure that green finance and industry not only invest in green technologies in developed countries but do so across different world regions.

The concluding chapter weaves together the threads of this complex story, drawing from it political, policy and institutional lessons. Given its policy-driven nature, Europe's energy transition will require an all-of-government approach to ensure both that the transition actually happens and that its social, economic and therefore political and geopolitical consequences are adequately addressed. This entails greater institutional synergies – if not mergers – between energy, environment and climate ministries or directorates, far greater attention to the energy transition in economic and finance institutions, and a greater coordination role played by chancelleries, bringing together governmental institutions, the private sector and civil society. It will also require a *primus inter pares* role of climate and energy in European foreign policy at both national and supranational levels, a Copernican revolution compared to the status quo, where these areas are often an afterthought in most foreign policy establishments. This, in turn, will require more funding and people dealing with climate and energy in foreign ministries as well as greater horizontal coordination between European capitals to socialize the laggards and develop synergies in climate diplomacy. A green Europe can represent a normative vision, an economic growth strategy and a route to a political Union only if it becomes a global Europe too.

1

A Green and Political Europe

The European Union is recognized within and beyond its borders as a global climate leader. At times, its practice has fallen short of its objectives, notably when its internal energy and international climate policies have proceeded along parallel and often non-communicating tracks. At other times, the EU's climate ambitions have failed to trigger positive change amongst other global powers, contributing to impasse in international climate talks. However, limits aside, the Union has undoubtedly played a pivotal role in global climate agenda setting, raising the level of international ambition and leading by policy example. This has been possible due to the relatively high degree of political convergence between Member States, alongside the fact that for many years climate policy fell at the margins of European domestic political debate.

Whilst joined at the hip with climate policy, European energy policy has proceeded along a different path. Marked by sharp differences between Member States, the latter retained control over their energy mixes and policies. Only with the entry into force of the Treaty of Lisbon in 2009, which coincided with the first gas crises between Russia and Ukraine, did the EU more visibly step into the energy field, mainly as an extension of its competences over environmental protection and the internal market, by and large with an aim to promote energy security.

After unpacking the parallel tracks through which EU climate and energy policies evolved over the years, this chapter explains how these paths have gradually converged during the last decade, especially with the Energy Union and then, more significantly, with the European Green Deal. Today, the EU's internal energy policies and global climate ambitions are intertwined and, as such, far more credible: energy security and the energy transition are increasingly seen as interlinked. While not necessarily being the best possible plan, nor one that is guaranteed to be implemented precisely as it was laid out, the European Green Deal represents the first attempt at the global level to chart a path to net zero greenhouse gas emissions by 2050, with a precise pathway to achieve the crucial 2030 milestone of a 55 per cent reduction of emissions. It is also understood as a plan that will increase the Union's energy security and autonomy, a strategic goal underlined dramatically by the Ukraine war. This is possible only through a greater synchronization of European climate and energy policies. For this effort, the EU must be credited.

The chapter then addresses the political significance of a green Europe. For the first time, climate and energy policy have moved from the margins to the beating heart of European integration, with the European Green Deal representing the EU's response to address the number one priority expressed by European citizens, first and foremost youth. As such, it constitutes the promise to transform a highly politicized Union into a political one that is more united, perceived as being more responsive to citizens' concerns, and therefore more legitimate in their eyes.

The EU as a global climate leader

The EU asserted itself as the global climate leader, overtaking the US amongst the developed economies, around the time of the 1992 Rio Earth Summit. It consolidated its climate leadership with the subsequent Kyoto Protocol in 1997, which committed the parties to internationally binding emission reduction targets, with developed countries accepting a heavier share of the burden under the principle of common but differentiated responsibilities (Oberthür and Roche Kelly 2008). When George W. Bush's administration withdrew from the Protocol in 2001, rejecting its binding targets, the EU remained a staunch defender of the pact, adamant on its implementation.

The growing convergence between Member States enabled the EU's global climate ambitions and spurred the Union to speak and act as one in international climate diplomacy in those years (Davis Cross 2018). As a Union of fifteen Member States until the 2004 eastern enlargement, European countries, while displaying vastly different energy policies, interests and mixes, increasingly recognized the anthropogenic nature of climate change and endorsed the need to counter it by acting together. As the body of climate science grew and consolidated, notably as presented in the successive reports of the Intergovernmental Panel on Climate Change (IPCC), the EU15 agreed to act together on climate change.

Climate change deniers of course existed in Europe back then, as they did elsewhere. We shall turn to this in chapter 2. However, unlike other countries, notably the US, European climate change deniers were few, being concentrated at the fringes of the political spectrum in different Member States. The broad recognition of climate change in Europe explained why climate policy was traditionally rather uncontroversial politically. The greater public acceptability of the role of the state in public policy also translated into an expectation that institutions should tackle the climate crisis, which can be read as the greatest market failure of all time. While often absent from the core of domestic political debates, revolving around jobs, growth, the welfare state, terrorism or migration, the need to address climate change was both accepted by all and ignored by most in the hustle and bustle of daily political life.

The broad agreement on climate change as well as its relatively low political salience in the 1990s explained why Member States, which retained full competence on energy policy, were happy to see the European Commission take the lead on international climate diplomacy. Entrusting EU institutions with the task of driving internationally Europe's climate ambitions made sense given the global public good nature of the challenge. Moreover, Member States did not have divergent national interests to press for in international climate talks and were happy to see the Commission step into the fray.

In the years that followed, the EU continued to speak and act as one, but divergences between Member States started creeping in, hampering the Union's effectiveness as a global climate leader (Davis Cross 2018). In particular, as the EU enlarged from fifteen to twenty-five and then twenty-seven and twenty-eight Member States, different degrees of climate consciousness came to the fore. Whereas northern, western and southern Member States broadly shared the same climate ambitions, some of the new members from central and eastern Europe did not. Hence, by the time of the 2009 15th Conference of the Parties in Copenhagen, whereas the EU continued to speak with one voice, its internal consensus was actually razor thin. The ensuing reluctance of EU institutions to rock the boat internally led to rigidities in the EU's external negotiating stance, which played no small role in the overall failure at Copenhagen. At COP15, the Union found itself unable to persuade other global players to follow its lead, whilst being unwilling to bargain up, down or sideways its own climate goals for fear of unravelling its fragile internal climate consensus.

After the global debacle in Copenhagen, the EU picked up the pieces (Oberthür 2011; Oberthür and Dupont 2021). It gradually worked on reconstituting an internal climate consensus, enabling it not only to remain at the forefront of the global climate agenda but also to push it through internationally. COP21 in Paris crowned this success. By 2015, the EU represented around 10 per cent of global emissions (Oberthür 2011). To ramp up global climate ambitions, it realized that working with partners was essential. Whereas the US–China agreement ahead of COP21 received far greater media attention and in many ways underpinned the Paris Agreement as discussed in chapter 4, the Agreement's level of ambition was ratcheted up by the 'high-ambition coalition' originally between the EU – and its twenty-eight Member States at the time – and seventy-nine African, Caribbean and Pacific countries. Only after the announcement of the coalition with this critical mass of countries signing up did other countries join in, notably the US, as well as Norway, Mexico, Colombia and many others.

Since the 2015 Paris Agreement, the EU has remained at the forefront of the global climate agenda. Its role is often not recognized nor sufficiently credited, and it certainly does not receive the same media attention compared to other international players, notably the US and China. Paradoxically this is precisely because of the Union's climate ambitions. Lying ahead of the global green curve, the major sticking points in international climate talks normally do not concern Europeans. It is not the EU that needs to be pressed to agree on more ambitious global climate targets and, while much remains to be done in terms of actual decarbonization, neither is it Europe that resists ambitious climate and clean energy policies. On the contrary, it is the EU which, out of the negotiating limelight, does much of the persuading and cajoling of others behind the scenes, leveraging its weight but above all its power of policy example. This has become increasingly true since the European Green Deal was unveiled, which linked organically European climate and energy policies, and propelled both at the heart of European politics and policy.

From European energy security to the European Green Deal

Despite climate and energy being joined at the hip, with the burning of fossil fuels representing three quarters of European greenhouse gas emissions, EU climate and energy policies have developed through different logics, dynamics and institutions up until the European Green Deal. This goes far in explaining why, despite the EU's climate ambitions and policies, the Union's own emissions reduction targets were often not met through (energy) practice. Unlike the case of climate, energy has traditionally been marked by different strategies, interests and mixes among Member States, as well as being governed by different institutions with very different networks and outlooks on the world. Different geographies, histories as well as economic and industrial interests and comparative advantages explain the different energy trajectories of Member States. Germany with its visceral political resistance to nuclear energy and highly developed automotive industry, Poland with its powerful coal sector, and France with its heavy dependence on nuclear energy are only a few examples of the wide variety of economic and energy contexts and interests within the EU. However, despite these differences, much like climate, energy policy has traditionally not made it amongst the hottest subjects in public and political debate. With the exception of a few countries in which pipeline politics has been at the forefront of domestic political debate, and apart from moments of energy price crisis, with their ensuing effects on jobs, industry and growth, energy policy at both Member State and European level has proceeded below the political radar.

Until the 2009 Treaty of Lisbon, energy policy fell essentially under the remit of Member States. Prior to Lisbon, the EU had passed legislation that directly and indirectly affected energy policy, including environmental directives on power plant emissions standards, the successive energy liberalization packages, the first security of supply directive, the various trans-European network initiatives as well as the growing body of internal market and competition discipline. However, energy policy, strictly speaking, fell under the exclusive competence of the Member States. This remains largely so to this day, with Article 194 of the Treaty on the Functioning of the EU reaffirming the principle of subsidiarity in the field of energy, and Member States remaining sovereign in determining their energy mixes and external supplies. Yet since the entry into force of the Lisbon Treaty, the EU has acquired limited competences in the energy field. These have been mainly an extension of its competences over the internal market, hence the heavy focus on market integration and the liberalization of network industries in the EU's approach to energy policy, particularly in the early post-Lisbon years.

To enter the energy field, the EU, notably the Commission, entrepreneurially transformed the proverbial crisis into an opportunity. The first Russia–Ukraine gas crises took place precisely at the time of the negotiation, ratification and entry into force of the Lisbon Treaty, and had a profound impact on the EU's energy outlook in those years. In 2006 – one year before the Lisbon Treaty was approved – and again in 2009 – the year it entered into force – Russia interrupted gas supplies to and through Ukraine, triggering critical shortages in the EU, notably in central and eastern Europe. Russia's use of its energy leverage towards Ukraine boomeranged politically within the EU, giving rise to a lively debate on European energy security, echoed and amplified across the Atlantic.

This boomerang effect was magnified by the fact that the EU and Russia were drifting apart both energy-wise and politically in those years. European and Russian energy models had started diverging, with rising resource nationalism in Russia, epitomized by the 2003 Yukos affair and Russia's refusal to ratify the Energy Charter Treaty.[3] On the part of the EU instead, the drive for energy liberalization and competition broke the monopoly–monopsony business model that had linked the Russian energy company Gazprom to individual European midstream operators. The EU and Russia were moving apart politically too. By the late 2000s, after the 2003–4 colour revolutions in Georgia and Ukraine, the Bucharest summit in which NATO declared that Georgia and Ukraine 'will become' members of the Alliance, and the 2008 Russia–Georgia war in South Ossetia, relations between Russia and the West tipped into an irreversible decline. In retrospect, Russian President Vladimir Putin's speech at the 2007 Munich Security Conference marked the turning point, putting an end to the dream of a cooperative Euro-Atlantic security space and a common market stretching from Lisbon to Vladivostok (Putin 2007).

This is not to deny that Member States had different sensitivities, interests and views with respect to Russia in those years. Until the 2014 Russian annexation of Crimea, some

Member States – notably in central and eastern Europe – openly blamed Russia, viewing it as a strategic threat, whilst others – in southern and western Europe – saw Moscow as a strategic partner. After 2014 and as a result of Russia's increasingly assertive foreign policies in Europe and the Middle East, Member State views converged on taking a tougher stance towards Moscow, particularly through sanctions. The EU's two-track approach featuring sanctions on the one hand whilst remaining open to selective engagement with Moscow on areas of common interest – including climate – reflected the EU's internal balance vis-à-vis Russia between 2014 and 2022.

In the first two decades of the twenty-first century, these hardening views towards Russia did not always translate into identical views on energy policy, however. Member States displayed vastly different dependences on Russian fossil fuels, ranging from a near totality in the case of the Baltic countries, Finland and Bulgaria, to around 10 per cent for the Netherlands or Romania (Elagina 2021). Furthermore, their political experiences were very different, with western and southern European countries considering Russia a responsible and reliable energy partner, while northern and eastern European countries were more sensitive to Russia's weaponization of energy. Indeed, until the 2022 war in Ukraine, Russia had politicized energy in relations with some countries but not others. Hence, some of the largest Member States like Germany and Italy sleepwalked into becoming more dependent on Russian gas in the 2010s, with over 40 per cent of their supplies coming from Russia by the time war broke out on the European continent.

Notwithstanding these differences, Member States agreed on the importance of strengthening the EU's energy security through a diversification of energy sources and routes. This push was partly driven and sustained by the US, especially in the early 2000s. In fact, the EU's major diversification project in those years – the mammoth Nabucco pipeline, which never came into being – was conceived in Washington DC. Given that much of this diversification would come geographically from or through the south – including the Caspian, Central

Asia as well as North Africa and the Eastern Mediterranean – southern Member States were happy to support this energy security agenda, notwithstanding their relatively more sympathetic views towards Moscow at the time.

The imperative of European energy security translated into law, institutions and policies in the 2010s. The Treaty of Lisbon, while reaffirming the sovereignty of Member States in determining their energy mixes, created a legal basis for the EU to act on energy policy (Article 194), notably regarding the development of the European energy market, the promotion of European energy interconnections and the security of energy supply. The High Representative was tasked with the role of promoting European energy security. In practice, this had several internal and international policy implications.

Within the EU, regulations were passed, infrastructure was built, and projects of common European interest started being directed more explicitly towards strengthening the internal resilience of European energy markets, security of supply and energy solidarity between Member States. Externally, the EU sought to diversify its energy sources and routes, notably regarding its gas imports, with policy-makers having primarily Russia front and centre of their minds. Hence, the grand plans for a southern gas corridor transporting Central Asian and Azerbaijani gas to Europe, the goal to significantly expand liquefied natural gas (LNG) capacities in the more vulnerable eastern European Member States, or the peace promise of Eastern Mediterranean gas, whose discoveries many hoped would end up in Europe (Baconi 2017). In the Eastern Mediterranean, the ideal solution both commercially and politically would have been to transport Israeli and Cypriot gas, and possibly also the far more significant reserves in Egypt, through Turkey and on to the EU. Part of such gas could have been consumed in Turkey, which was growing demographically and above all economically at an astounding pace in the early 2000s. The rest could have been connected to the Turkish pipeline network, which already received gas from the Caucasus, and, through the southern gas corridor, these supplies would have reached the EU.

Some of these initiatives came to fruition. Internal regulatory work proceeded, for instance with the security of gas supply regulation in 2017, and the third energy package and the clean energy package in 2019. European energy infrastructure became more resilient and interconnected. Other initiatives instead shrivelled to their former selves. The southern gas corridor was never realized as originally planned. With Russia consolidating its strategic grip on Central Asia and China's energy demand growing apace, Central Asian and Caspian gas has not made its way to Europe but has rather headed eastwards. The only real exception has been Azerbaijan, whose gas eventually reached the EU in 2020 after a fraught 'not in my backyard' (NIMBY) battle in southern Italy. Yet the Trans Adriatic Pipeline (TAP) carries an underwhelming 10 billion cubic metres (bcm) of Azerbaijani gas annually, a far cry from the original grandiose European plans for a southern gas corridor carrying 50 bcm annually, that would have significantly reduced European gas dependence on Russia (Morrison 2017).

Dampening these hopes further was the sharp deterioration of EU–Turkey relations and the heightened crisis in the Eastern Mediterranean in the 2010s. Rather than acting as a catalyst for peace, gas discoveries in Cyprus, Israel, Egypt and Lebanon have been trapped within the conflictual geopolitics of the region. Eastern Mediterranean reserves are unlikely to be piped to Europe. They are far more likely to be sold to domestic markets or liquefied and sold on global gas markets, paradoxically including Turkey too.

What is striking about the EU energy policy debate in the 2000s and 2010s is that it unfolded following a seemingly disconnected logic from the consolidating climate consciousness within the EU in that period. This plays no small role in explaining why in those years the EU struggled to meet the ambitious decarbonization targets that it set out for itself. There were, of course, points of contact between energy and climate. The 2007 '20-20-20' energy and climate package, which aimed to reach a 20 per cent cut in greenhouse gas emissions, 20 per cent energy from renewable sources, and a

20 per cent increase in energy efficiency by 2020 compared to 1990 levels, clearly tied energy and climate policies together. So did the establishment of the Emissions Trading Scheme (ETS) in 2009, a cap-and-trade system for heat and power generation, energy-intensive industry and domestic aviation, covering collectively 40 per cent of European emissions. As these targets were progressively upgraded over the following decade, with the 2014 goal to reduce emissions by 40 per cent by 2030 and at least by 80 per cent by 2050 and the ensuing revision of renewables and energy efficiency targets, the linkage between climate and energy became tighter. Furthermore, the Energy Union, one of the flagships of Jean-Claude Juncker's Commission between 2014 and 2019, systematized and accelerated the work to bring together the European climate and energy policy debates. Energy efficiency, decarbonization, and research, innovation and competitiveness in low-carbon and green technologies were three of the five pillars of the Energy Union, alongside energy security and energy market integration.

However, while brought together under the same policy framework, the climate and energy security agendas were broadly understood as separate priorities rather than being intertwined (Strambo et al. 2015). The former was centred on carbon pricing, energy efficiency and the promotion of renewables. The latter, while being furthered also through the promotion of energy efficiency and home-grown renewable energy sources, revolved largely around the increase of fossil fuels imports from a diversified set of suppliers and routes, as well as closer integration between European energy markets. Furthermore, if one had to point to an unspoken political hierarchy between these two sets of goals, energy security clearly had the upper hand. Institutionally, for instance, whereas the Energy Union Commissioner, at the time Maroš Šefčovič, was also Vice President of the Commission, Miguel Arias Cañete was 'only' the Commissioner responsible for Climate Action and Energy.

By the mid-2010s, the context changed. On the one hand, the 2015 Paris Agreement propelled climate action to an entirely new level of global ambition. On the other hand, energy prices dropped in 2014, reducing the political salience of energy security in those years. Amidst this changed context, the European Commission, led by Ursula von der Leyen, turned the implicit institutional and political hierarchy between energy and climate on its head. The European Green Deal was always meant to be the signature initiative of von der Leyen's Commission, being presented as such when first unveiled in December 2019 (European Commission 2019). Her Commission reflected institutionally this green push, with its first Vice President, Frans Timmermans, being responsible for seeing the European Green Deal through. Only a few months later, the announcement of the European Green Deal, the pandemic, the EU's economic response to it, and the growing public awareness of climate change propelled a green Europe to new heights. The European Green Deal began its journey from being aspirational to becoming a reality.

Through the European Green Deal, climate policy, which has been under the remit of the Commission since the late 1990s, started affecting ever more intimately energy policy, which still falls mainly under the competence of Member States. This represents the best guarantee that the EU's climate targets will eventually be met. In fact, one could go as far as claiming that EU climate policy started driving European energy policies. Especially in times of low energy prices up until 2021, climate policy entirely replaced energy security and market integration as the dominant policy paradigms and goals.

Since the announcement of the European Green Deal, the EU has equipped itself with a Climate Law, enshrining in legislation its political pledge to reach climate neutrality by 2050, requesting that all EU policies contribute to reaching the European Green Deal's objectives (European Commission 2021b). The EU has committed to reducing emissions by at least 55 per cent by 2030 compared to 1990 levels. It has earmarked one third of its €1.8 trillion seven-year budget for the transition. Within this, 40 per cent of Member States' projects funded by the EU's post-pandemic recovery and resilience funds must be green. At the same time, the European Investment Bank will support €1 trillion of investments for climate action by 2030. To reach its 2030 emissions reduction milestone, the Commission put forward the 'Fit for 55' package, advancing thirteen proposals for new or revised targets,[4] regulations and taxation measures across a range of sectors, from sustainable transport, to energy efficiency, carbon pricing, industrial policy, renewable investments and biodiversity (European Commission 2021c). For the first time, the EU is advancing an all-of-government ecological approach, featuring the prime goal of decarbonization, while covering broader environmental matters too, from biodiversity, forestry, pollution, food, land use as well as the European holy grail of agriculture. Taken together, the European Green Deal encapsulates the EU's vision for a sustainable, fair and prosperous society.

The European Green Deal – and the decarbonization goal that lies at its core – was portrayed as the way to achieve not 'only' the EU's 'new' climate goal but also the 'old' objective of energy security. In the first two years of the Green Deal, this came with a hefty dose of idealism. Energy security was a thing of the past, which would painlessly come about through decarbonization. In European energy and climate mainstream talk, the very word 'diversification' was expunged from the accepted vocabulary insofar as it was associated with fossil fuels. The 2021 energy price spike and the 2022 Ukraine war forced back a degree of pragmatism. But the return of the energy security agenda, including through diversified fossil relationships, rather than replacing the green agenda, became embedded in it. Today, the vision of a green Europe has become the dominant narrative driving European integration, enveloping all others. Specifically, the European Green Deal has transformed the energy security debate compared to how it was first conceived in the early years of this century. Inevitably, energy security requires diversified fossil suppliers and routes through energy diplomacy and investments in the short term. The Ukraine war was a rude reminder of this urgency. However, in the medium and long terms, energy security would be achieved also and mainly by reducing the reliance on fossil sources, replacing these with green capacities. The extent and pace at which the 'diversified' short term can tip into the 'decarbonized' medium and long term is a complex subject, which chapter 2 will explore. The point here is that in the European policy debate, the green agenda has merged with the energy security one, which was so prominent in the 2000s, gradually faded in the 2010s and made a dramatic comeback in the 2020s.

At the same time, the energy and climate debates, now fused in one, rather than remaining at the margins of European political debate and practice, have been propelled into the spotlight. By becoming more strategically and politically salient, EU climate and energy policies have created an unprecedented opportunity for the EU to put behind itself decades in which it was seen as distant, irrelevant, unresponsive and illegitimate in the eyes of many European citizens. As the EU's output legitimacy was tarnished – notably as regards growth, jobs, social justice and migration – the structural limits of its input and throughput legitimacy came to the fore (Schmidt 2013). The European project, increasingly seen as a neoliberal or ordo-liberal project, became associated with socio-economic disparities, restrictions and injustices. Running into the same critiques as globalization, the bureaucratic and cosmopolitan elite in 'Brussels' became the bogeyman responsible for most European maladies, rather than representing the political and institutional crystallization of peace and prosperity on the European continent.

A green Europe, if associated both with climate action *and* with energy security, growth, jobs and social justice, offers a political route out of this impasse. If the EU pursues its net zero goal in a manner that fosters security and employment, reduces disparities and promotes industry, it can realize its normative vision, protect citizens and businesses, and narrow the economic gap that was generated when manufacturing migrated east and widened as the digital race started being lost to the US and China. Were it to succeed, the EU would be well on its way towards becoming a political and therefore more united, responsive and legitimate Union.

The political rationale underpinning
the European Green Deal

The vision of a green Europe is exactly right. The climate crisis *is* an existential crisis. Never has humanity experienced an almost doubling of CO_2 atmospheric concentration, with the related increase in global average temperatures, as it is today. Sea level rise, melting ice sheets and extreme weather events, with the ensuing security, social and economic consequences we are experiencing are only a mild beginning of what lies ahead. The IPCC (2021) highlighted how greenhouse gas emissions have already led to a 1.1°C increase in global average temperatures since pre-industrial times, with anthropogenic warming expected to reach or exceed 1.5°C over the next two decades. The Earth has seen this before, with the Paleocene–Eocene Thermal Maximum being the closest parallel to the anthropogenic warming we are inflicting upon our planet now.[5] That was 55 million years ago, and all the science suggests both that the Earth will eventually adapt, and that it will take tens of thousands of years to do so (Archer 2016). In other words, the planet will survive, human civilization, as we know it, may not.

European citizens get it. A European Investment Bank survey revealed how the Covid-19 pandemic heightened the public's awareness of the climate crisis and strengthened the belief that economic recovery and the energy transition must go hand in hand (EIB 2021a). In particular, young Europeans care about climate change and sustainability more than any other public policy issue. Their mobilization – notably Fridays for Future – has been a prime driver behind the EU's stepped-up commitment to addressing climate. An Ipsos (2021) survey found that younger Europeans, despite the pandemic and the economic crisis, still place global warming and environmental degradation as their top two priorities, seeing these as affecting deeply their future lives and jobs. And it is precisely the younger generations that the EU must engage if it is to have a political future: the EU's fate hinges on being recognized as a central piece of the solution to the climate crisis, igniting passion in the public by acting on what citizens are passionate about. This makes the European Green Deal and its success pivotal both in itself and for the political future of Europe.

It is striking, in fact, to see how climate and energy have rapidly risen as prime subjects of political debate in Europe. As the effects of climate change are felt by the public, notably through heat waves, intense rainfall, floods and melting glaciers, politics has responded. Already in the context of the 2019 European Parliament elections, climate had risen up the political agenda, with the European Climate Foundation (2019) highlighting that more than three quarters of European voters expected their representatives to tackle climate change. Since then, public attention has increased exponentially. Post-pandemic elections in European countries, notably Germany, but also European Economic Area member Norway, revealed how much climate and energy have climbed up the political ladder. Whereas, until 2019, climate was one amongst several concerns of the German public, against the backdrop of the severe flooding that killed 180 people in the summer of 2021, it overtook the Covid-19 pandemic as the number one issue for the public, occupying centre stage in the general election campaign that year (*The Economist* 2021). In the Netherlands too, citizens identified climate and energy as increasingly salient issues, key in shaping their voting preferences in the 2021 election (Nieuwenhout 2021).

The rising political salience of climate and energy does not mean that these subjects are set to remain politically uncontroversial, however. The climate consensus may in fact be rather 'soft' and fragile (Fieschi 2022). As the relevance of climate change rises, the political divisiveness, particularly around the pace and shape of the energy transition, will likely increase too, marking a sharp difference from the decades in which acting on climate change was politically consensual but also on the margins of domestic political debate in Europe. As climate and energy policy become intertwined and thus more concrete, difficult political choices and policy trade-offs will need to be made. Lifestyles will need to change, causing support for and resistance against the transition. Economic winners and losers will make their voices heard, making the transition both more politically salient and infinitely more contentious in Europe.

To the extent that the EU defines itself through the green agenda, it will become increasingly enmeshed and drawn into those domestic political debates: the EU will become ever more politicized. This is part of a trend. The politicization of the EU really began with the Eurozone crisis, as debates about austerity were front and centre of domestic politics in northern and southern Member States alike (Dokos et al. 2013). The EU's politicization persisted over the migration crisis, from the controversial agreement with Turkey on refugees to the acrimonious debate over (the lack of) European solidarity between southern, northern and eastern European countries (Okyay 2019). However, in both these cases, the EU's action was either deficient or lacking altogether, in one case scraping through the crisis and in the other being almost entirely missing in action. Europe, and in particular 'Brussels', was associated with passivity, injustice and division. This meant that the politicization of the EU, rather than opening the way to a political Union, became the prime obstacle to it, with nationalist forces galvanizing domestic support by lashing out at Europe. The EU's poor performance on the economy and migration generated a wide gap between the reality of a politicized Union and the vision of a political, and thus more integrated, accountable and legitimate, Union.

A green Europe promises a different way. The vision of a green EU is triggering public passion and domestic political interest: the Union will continue being politicized. However, if that vision is realized, it can also open the way to a more tightly knit political, and not just a politicized, Union. Three reasons underpin this claim.

First, to the extent that the EU works and delivers on the first public priority of the European public, 'Brussels' can cease being viewed as distant and useless at best, selfish and damaging at worst. If the Union delivers on climate, energy security and green growth, its output legitimacy will consequently grow. Second, up until when European countries share – albeit to different degrees – the belief that they need to act on climate change, the green agenda, which has been propelled to the core of the European project, can foster unity between Member States and that unity will underpin the EU's standing in the world. That window of opportunity may not last forever, but it does represent an opportunity to bridge the divides between European countries that emerged and worsened over the last decades, and reconstitute the EU's role and reputation in the world. Fiscal policy, migration, Russia and rule of law have been amongst the most salient questions tearing the Union apart politically. The shared consensus to act on climate change and energy security could catalyse greater political unity between them. Finally, the EU's green agenda has the potential to be associated not simply with bans, restrictions and constraints: it could become an economic opportunity and therefore be connected to the generation of jobs, income and growth. This potentially sets the green agenda apart from the policy themes which the EU has been mainly associated with in recent decades such as the euro, austerity or migration. In these cases, the EU was the bad cop: in citizens' minds, it came to represent a detached and bureaucratic ivory tower that mechanically set the rules, inhumanely policed their compliance, and demonstrated perennial division, hypocrisy and lack of solidarity. By contrast, the green agenda should be about security rather than dependence, production rather than constraint, about spending rather than saving, it should be social rather than ordo-liberal, and above all, it should be about navigating one of the greatest challenges of our time. This is the vision and the promise. The challenge now lies in translating it into practice.

2

A Green Europe and the Future of Liberal Democracy

Liberal democracy has been under attack for the last decade. The US neoconservative agenda, far from strengthening the global appeal of liberal democratic values, undermined their credibility in many parts of the world. The second Iraq war, conducted under false pretences, was particularly damaging in that respect. Thereafter, the global financial crisis, triggered by and exposing the huge excesses of Wall Street, weakened the American neoliberal economic model, just at the time when Chinese growth was being measured in double digits lifting hundreds of millions out of poverty (Derviş and Tocci 2022). The global financial crisis and the ensuing European sovereign debt crisis, alongside the rise of multipolarity and the return of geostrategic rivalry, were read as evidence of a relative decline of the democratic West, and the rise of powers that either openly defied liberal democratic values or just refuted the ideological and material hegemony of the US (Cooley and Nexon 2021). The far from stellar outcomes of the colour revolutions in eastern Europe, not to mention the Arab uprisings in North Africa and the Middle East, pointed to an entrenchment of authoritarianism (Diamond 2015). And the surge of nationalist populism in the West, including the election of Donald Trump, the Brexit refer-endum, the authoritarian bents of the Polish and Hungarian governments, and the growth of Eurosceptic parties across the EU, reflected a crisis of democracy at the very core of the liberal West.[6]

The Covid-19 pandemic has halted, at least temporarily, the nationalist populist wave sweeping across Europe. As a result, these Eurosceptic movements are on the lookout for a new story to tell. The energy transition, if ill managed, could provide them with powerful new ammunition. This would endanger a green Europe, the revival of the European project, and the future of liberal democracy all at the same time. After outlining the nationalist populist threat to liberal democracy and the EU, this chapter highlights the potentially regressive distributional effects of an ill-managed energy transition and how these could provide new ammunition to Eurosceptic forces. It zooms in on the 2021–2 energy crisis, drawing its lessons for the bumpy transition ahead. It concludes by examining the policy measures that are being implemented to mitigate the adverse socio-economic effects of the transition, and the further measures that could be developed to fully rise to the socio-economic and therefore political challenge of a green Europe.

The internal and external opponents of liberal democracy, and of the European Union

The attack on liberal democracy over the last years has come both from within and from outside, with authoritarian countries and Western nationalist populist parties drawing on similar political and policy narratives. These internal and external opponents of liberal democracy have at times supported each other explicitly, be it financially or politically. The political and financial ties between the Kremlin and extreme right-wing parties such as the French National Rally and the Italian League or Brothers of Italy are cases in point. More often, this support has been implicit, taking essentially the form of mutually reinforcing narratives.

These narratives have sought to make the case that liberal democracy does not deliver. Authoritarian leaders like Russian President Putin have explicitly emphasized the inability of liberal democratic systems to provide effectively for the security and wellbeing of their citizens (Tiounine and Hannen 2019). Distracted by a hegemonic drive for regime change around the world, the Russian president claims that the West has become victim of its own moral, political, institutional and socio-economic fragilities. The EU's sovereign debt crisis, terrorism, Brexit, the refugee crisis and the early phases of the Covid-19 pandemic were used as evidence for Russia to make its case of an imminent collapse of the liberal European project.

China, more indirectly, has been putting forward a similar argument. Today's global confrontation, while playing out primarily in the economic and technological domains, at its heart is a competition between political systems and ideologies. It underpins China's implicit claim that its (autocratic) model of governance is superior to liberal democracy in terms of economic growth, use of technology, security or public health (Derviş and Tocci 2022). China does not necessarily want to export its model of governance, but it does need to continue demonstrating to its own people that its political system delivers, ideally more and better than liberal democracies do.

Authoritarian regimes beyond the West and illiberal nationalist populist leaders within Western liberal democracies often voice similar political and policy views. Like authoritarian regimes, European nationalist populists have also attacked the moral corruption of the liberal West, its loss of traditional values, and the failure of technocratic liberal governance to deliver fair policy outcomes as well as to reflect and respond to the interests of the 'people'. They have then exaggerated and distorted these failures through misinformation and disinformation campaigns, whilst explicitly or implicitly advocating different models of governance. The arguments these opponents have raised have a common thread: they draw on and seek to galvanize the socio-economic grievances, and identity and demographic fears of the left behind, whilst claiming that the cosmopolitan, technocratic and liberal elite has pursued its political agenda at their expense.

Epitomizing technocratic, liberal and cosmopolitan governance is, of course, the European Union itself. No surprise that within the EU, nationalist populist political forces and leaders are also, invariably, Eurosceptic. This does not mean that all Eurosceptics are automatically illiberal or undemocratic. But illiberal, nationalist and undemocratic forces in Europe have generally been Eurosceptic. In fact, given the quintessentially liberal nature of the European project, whereas a nationalist and illiberal Europe is possible, a nationalist and illiberal EU is not. Indeed, the challenge to liberal democracy within the EU, especially between the global financial crisis and the outbreak of Covid-19, came alongside a rise and consolidation of Euroscepticism across the EU at around 50 per cent of the electorate, peaking at almost 70 per cent in the mid-2010s.

Picking on the fragilities of liberal democracies

The policy platforms that Eurosceptic forces have put forth have resonated amongst the public to the extent they have successfully intercepted, while manipulating, fomenting and distorting, elements of truth. It is true that the global financial crisis and the Eurozone crisis that ensued laid bare and worsened the excesses of neoliberalism and hyper-globalization, deepening socio-economic inequalities and impoverishing the middle classes (Piketty 2014). It is true that the perception of ungoverned migration has played into the identity fears of demographically shrinking societies and the socio-economic insecurities of the lower middle classes (Dempsey 2018).

And it is certainly true that on both counts the EU did not rise to the challenge. The EU scraped through the Eurozone crisis. It did 'whatever it takes', to cite then European Central Bank President Mario Draghi, to save the euro and therefore the Union, putting in place the first building blocks of a banking union. But having taken a step back from the brink, it left the banking union incomplete, without doing much else. The four presidents' report in 2012, which laid down the journey from a banking to a fiscal and a political union, ended up collecting dust on policy-makers' shelves, failing to garner consensus between Member States (European Council 2012). Most significantly, even when the Eurozone crisis gave way to a renewed period of (modest) economic growth, little was done to address and reduce the socio-economic, geographic and gender disparities that have been growing within the Union. Despite all the talk of a social Europe, in practice not much happened (Vandenbroucke et al. 2017). While welfare systems across Member States were being weakened and hollowed out, the supranational EU level, lacking the competences to forcefully step into the social arena and unable to shed its (neo)liberal ethos, talked the talk without walking the walk of a social Europe.

The EU has had even less to show for its efforts on migration. While seeking to strengthen its external migration partnerships and border management policies, the EU has failed to move forward on a common migration and asylum policy. This means that the legal resettlement of refugees from third countries and their internal relocation between Member States did not see the light of day. Only with the Ukraine war in 2022 did the EU activate for the first time its temporary protection directive. Not to mention EU labour migration policy, which has had little impact, and therefore has been entirely absent as a dimension in the EU's migration partnerships with countries of origin and transit. The result has been the partial reduction of migration and refugee flows through strengthened border controls, and security cooperation with and economic assistance to transit and origin countries. But the profound lack of solidarity within the Union and the ease with which the migration card can be leveraged by third countries to extract benefits from the EU indicate a profound European vulnerability. One only needs to think of the attempt by Belarusian autocrat Alexander Lukashenko to weaponize migration vis-à-vis the EU in the autumn of 2021 to highlight the point.

Authoritarian regimes and Western nationalist forces have then magnified and distorted these weaknesses through increasingly sophisticated misinformation and disinformation campaigns. The policy responses they have put forward have not held water, as nationalist Eurosceptics may share the same political values, but precisely because of it, they are unable to muster shared policy platforms, which would require compromising on their nationalist agendas. Italian, French, German and Hungarian nationalists may all dislike migrants and deeper economic integration, but this automatically means they are unable to agree on whether and how to reform the Eurozone or how to relocate and resettle refugees to cite the most obvious examples.

However, nationalists' inability to deliver policy outcomes has not prevented them from galvanizing popular support. Eurosceptic parties have consolidated just under a third of the seats in the European Parliament (Treib 2021), and they have either won – Poland, Hungary, Slovenia – or have come dangerously close – France, Italy – to reaching prolonged executive power.

With the Eurozone and the migration crises slowly fading, and Brexit certainly not being read as a stunning political success in continental Europe, Eurosceptic political forces need new fuel to fan the flames of their public support: they are in search of a new political banner. The Covid-19 pandemic initially appeared to represent such an opportunity. In early 2020, when the EU was initially slow to react, then in early 2021 when its vaccination rollout intertwined with the acrimonious post-Brexit EU–UK relationship, and finally as EU Member States made vaccines mandatory for work, leisure and travel, some Eurosceptics in opposition flirted with fringe views on Covid-19 (Spilimbergo 2021). Some left-wing populists blamed public health systems, hollowed out by decades of neoliberalism. Right-wing populists lashed out against globalism, highlighting the 'foreign' nature of the threat, as well as the encroachment of public authorities on citizens' freedoms (Wondreys and Mudde 2020). Authoritarian populists toyed with macho politics, minimizing the health crisis and the social measures to contain it.

However, unlike the US, where Covid-19 got entangled in the ever more divisive domestic political scene, European populists failed to exploit successfully the pandemic crisis (Bobba and Hubé 2021). The evidence is mixed, and there were demonstrations against lockdowns, restrictions and vaccinations across several European countries. However, while making much noise (Jones et al. 2021), the numbers were relatively contained. Overall the evidence suggests that the pandemic has halted, at least temporarily, what appeared to be an unstoppable wave of Euroscepticism sweeping across the Union (Betz 2020). National election results in the Netherlands, Germany, the Czech Republic and Slovenia in 2021–2 confirmed this trend. The reasons for this are difficult to discern. They may range from the drama of the pandemic that drove a public demand for reliable, responsible and moderate government, to a return of public confidence in facts, competence and science. The causes are disputed, but the correlation is there: the pandemic did not work well politically for Eurosceptics.

The temporary arrest of Eurosceptic nationalism may also have to do with the fact that compared to previous crises, the EU navigated this one incomparably better. The EU initially struggled to coordinate the chaotic sequence of closures, restrictions and lockdowns, but it eventually put a system in place both for internal movements within the EU and international ones with the rest of the world. Its vaccination rollout was slow to start but it eventually paid off, particularly when compared to other developed countries like the US. The decision to coordinate the procurement and distribution of vaccines at European level ensured that the doses arrived not only in the larger and wealthier Member States but also in the smaller and poorer ones. Perhaps most significantly, the EU reacted economically to the pandemic crisis by putting behind itself years of economic austerity. Alongside the injection of liquidity through the European Central Bank temporary asset purchase programme, the suspension of the Growth and Stability Pact on debit, deficit and inflation, and the establishment of a European unemployment scheme, the EU agreed on a €750 billion NextGenerationEU Fund, embedded in the most ambitious seven-year budget in the history of the European project.

While this has not yet resulted in Member States agreeing to mutualize their national debts, the EU's response to the pandemic de facto entailed a significant transfer of funds from stronger to more fragile European economies. It is Member States like Italy and Spain that received the lion's share of these EU funds. Furthermore, the EU's budget holds within it the seeds for the EU to generate its own fiscal resources, through measures such as the reallocation of taxing rights on multinationals as agreed by the OECD and, as we shall see below, the extended Emissions Trading Scheme and the Carbon Border Adjustment Mechanism. In other words, the magic word that lies at the core of the European integration project – solidarity – that had gone missing over more than a decade, seems to have been found once again.

This has two implications. The first is that Eurosceptics are still on the lookout for a new rallying cry: they need a new banner to mobilize the public and attack the EU for its failures, hypocrisy and aloofness. Second, it suggests that when the EU rises to the challenge posed by a crisis, it sucks wind out of the sails of Eurosceptic politics: output legitimacy matters (Schmidt 2013). Both implications are not easily proved empirically but they do hold water analytically and politically. As such, they point to the risk but also to the opportunity that lies behind the link between the energy transition and the future of liberal democracy in Europe and beyond.

The distributional effects of the energy transition

If badly managed, the socio-economic consequences of the energy transition could represent a political risk both to the green agenda and to liberal democracy. Were it to aggravate the already deep socio-economic disparities and hollow out European industry, the transition would understandably fuel public grievance and discontent. This could become a weapon in the hands of nationalist forces in Europe, as much as authoritarian regimes in their ideological competition with the liberal democratic West.

The EU narrative has emphasized that its green agenda is also its growth strategy: the growth of GDP and CO_2 are decoupled. It is certainly true that the technological and industrial transformation that underpins the energy transition can and should represent a powerful engine for growth and that a climate-neutral Europe can also be a more prosperous one. In fact, proponents of a green Europe would argue that economic development, rather than being a positive side-effect of decarbonization, is a prime driver of it.

However, this does not mean automatically that the transition itself may not have adverse distributional effects, featuring deepening rural–urban cleavages, job destruction in carbon-intensive sectors, a loss of competitiveness of European industry as well as rising energy costs for vulnerable individuals and small and medium-sized enterprises. This is all the more acute if the transition takes place in the midst of an energy crisis, with repercussions for many years to come. If these adverse distributional effects were to materialize, they could represent a powerful argument both against the transition and against the EU and liberal democracies promoting it.

A growing body of literature discusses why and how, in light of existing disparities, the energy transition could be deeply regressive (Zachmann et al. 2018). At a macro level, if it is true that the energy transition is one that will shift wealth further away from labour and towards capital, then this could reinforce the entrenched trend of growing inequalities (Luciani 2020). Added to this, if this transition is both technology and policy driven as discussed in the introduction, the policy instruments deployed to accelerate it – from carbon taxes, cap-and-trade schemes, to regulation – could add fuel to the fire. Insofar as energy expenditure represents a relatively higher share of lower-income households' budgets, not to mention the indirect effect that energy costs have on consumer prices, the rising cost of energy that will accompany the transition could fall disproportionally onto their shoulders. Evidence from Europe confirms this, with countries with relatively lower income levels and poorer building insulation being more exposed (European

Energy Poverty Advisory Hub 2021). Worse still, lower-income individuals, particularly those living in rural areas, are unlikely to shield themselves against those rising costs, given their reliance on private transport, and their inability to afford electric vehicles or better insulate their homes. This would exacerbate both socio-economic inequalities and rural–urban divides (Falchetta and Noussan 2021). The same argument applies to business, with small and medium-sized enterprises and companies with narrower financial buffers being more exposed to the higher and more volatile costs that are likely to accompany the energy transition.

The 2021–2 energy price spike as a canary in the transition coalmine

The 2021–2 energy crisis, in which wholesale electricity prices soared, largely driven by an exponential increase in wholesale gas prices, can be read as a canary in the coalmine. The onset of the crisis is broadly explained by market forces (European Commission 2021b; IEA 2021b; Tagliapietra and Zachmann 2021). With the reopening of the economy after lockdowns and unprecedented fiscal stimuli across the Union, European energy demand picked up. Increased demand in residential heating due to a cold winter and remote working, coupled with greater demand from power generation and industry due to economic recovery, led to a growth in energy demand.

Supply did not keep pace when this happened. Renewables today represent just over a third of the sources of European electricity generation, but lag well behind in other sectors such as heating/cooling and transport (Bianchi and Colantoni 2021). In addition, the weather did not come to the rescue, with a lower than average wind availability in the summer of 2021. The demand for fossil sources thus shot up. With carbon prices having risen exponentially over the course of the pandemic, reaching almost €90 per ton of CO_2 by the end of 2021, the demand for less carbon-intensive gas over coal rose.[7] However, coming after two price drops, in 2014 and

2020, investments in oil and gas almost halved compared to pre-2014 levels, with international investors reluctant to put money into what looked like a dying business. Furthermore, with demand increasing not only in Europe, but also in Asia – notably China, Japan and South Korea – much of the limited LNG supplies in increasingly globalized gas markets travelled to the more profitable east.

Within this largely market-driven story, Russia entered the scene. In fairness, at the beginning of the price spike, Moscow was not the main culprit of rising energy prices. It respected its long-term gas contracts, although it did not rush to rescue Europe by increasing supplies on spot markets on which the EU increasingly relied. This was explained partly by its own growing domestic demand and delayed infrastructure maintenance, and partly by a dose of political gloating and arm-twisting after years in which low commodity prices, and Europe's sagging demand and prospected transition strengthened the latter's bargaining hand over Moscow (Reznik and Meyer 2021). This was all the more so, given that the Nord Stream II gas pipeline, essentially complete, had not received authorization by German and EU regulatory authorities at the time.

However, rising energy prices certainly did provide a propitious strategic environment for Putin to build up tension over Ukraine by amassing almost 200,000 troops on the Ukrainian border. As autumn tipped into winter, Russia probably deliberately began weaponizing energy, by reducing its gas storage levels in Europe, which led to a further rise in gas prices. With tensions escalating into a full-blown invasion, energy became a prime feature of the confrontation. The EU and the US responded through severe sanctions, including Germany's suspension of Nord Stream II's certification, the US energy embargo, the phased EU coal and oil embargo, as well as the suspension of Russia from the SWIFT payment system. Russia too flexed its energy muscle towards Europe, reducing or interrupting gas supplies towards some Member States. In short, a market-driven energy crisis in 2021 morphed into and was made worse as a result of geopolitics in 2022.

For all these reasons, European gas and ensuing electricity prices spiked to record highs, awakening the spectre of stagflation. This has led to an impossibly difficult predicament for carbon-intensive industries in Europe, from steel, cement and aluminium, to paper, fertilizers and refineries amongst many more, caught between the Scylla of soaring commodity prices and the Charybdis of the ever-growing cost of CO_2. It also raised the concern, notably in southern and eastern European countries, that the rise in energy prices would generate unsustainably high costs for lower- and middle-income households, already struggling to recover from the economic consequences of the pandemic. Already in pre-crisis times in 2019, around 7 per cent of the population across the EU – 31 million people – were unable to keep homes adequately warm, with large variations between Member States and income groups. The 2021–2 spike in electricity and gas prices made this number rise dramatically (European Commission 2021b). The costs weighing disproportionally on lower-income households are not limited to utility bills and transport. They affect food supplies and prices as well. This is because natural gas accounts for around 80 per cent of the variable costs of fertilizer components such as ammonia. With fertilizer prices soaring in response to the gas price rise, compounded by supply interruptions caused by the Ukraine war, food insecurity rose to record highs.

The relationship between the energy price spike and the transition is clearly complex. On one level, it is true that the crisis is the consequence of insufficient transition to date (IEA 2021a; European Commission 2021b). For the sake of argument, if fossil sources were entirely absent from electricity generation and the penetration of renewables – with the necessary storage capacities – were complete, gas would not represent the marginal source determining electricity prices in Europe, and gas and electricity prices would be decoupled. It is also true that, drawing from the current supply crunch the lesson that investments in oil and gas should continue unabated is like saying that faced with withdrawal symptoms, the addict should be given another dose. In fact, it is unlikely, if not impossible, that investments in the oil and gas sector will resume to their pre-2014 price drop levels. Private capital in Europe is increasingly driven to invest in green technologies rather than fossil fuels. Moreover, European energy companies,[8] in the search for a continued future, are redirecting investments into their own decarbonization. Whereas oil and gas investments will continue in the decades ahead, it is highly unlikely that they will revert to the levels seen in past decades.

However, precisely because of this, we can expect that the energy transition, akin to a revolution in view of its speed, depth and breadth, will be accompanied by a greater structural volatility in prices. As fossil fuel supplies increasingly respond not only to energy demand but also to climate policy and public opinion, energy prices are likely to become more volatile in the (long) period of transition away from hydrocarbons. Public policy needs to address the consequences this might have, finding the right balance in the delicate trilemma of assuring energy security, affordability and climate sustainability, rather than halt or even slow down the pace of change.

In navigating this trilemma and seeking a balance over the course of a bumpy transition, the 2021–2 energy crisis in Europe points to two lessons. First, that the transition, and the volatility embedded in it, will be costly, and that cost could end up hampering growth, harming industrial competitiveness and exacerbating socio-economic disparities (Tagliapietra and Zachmann 2021). Commodity markets have often witnessed super-cycles, with the volatility that has come with it. However, added to this structural volatility is the intrinsic uncertainty that accompanies the transition, with ensuing risks of under- (and over-) investments along the way. The same is true of oil and gas prices. So long as fossil sources represent a share of electricity generation – i.e., for many years to come – it is true of electricity prices as well, unless electricity pricing systems are rethought altogether.

The second is that whereas CO_2 prices represent only a small part of the story behind the 2021–2 price rise (European Commission 2021b),[9] if those prices continue to rise – as indeed they should to make the transition happen – they will come to represent a growing share of the cost of fossil sources, notably coal, but also oil and gas. In other words, if 2022's €70–100 per ton of CO_2 represents only a fraction of the explanation of the current energy price rise, were that price to continue growing, reaching €120–140 as suggested by the IEA's net zero scenario, the weight of carbon pricing in overall energy prices would increase too (IEA 2021a). This could weigh on the shoulders of European carbon-intensive industries that would lose their competitiveness and on all those individuals who cannot afford low-carbon alternatives. And it is not just necessary for carbon prices to rise to make the transition happen; it is also likely that they will rise as the market internalizes the irreversibility of change. Again, the dynamic underpinning the winter of 2021–2 is instructive, indicating a two-way street in which both fossil fuel *and* CO_2 prices have concomitantly risen.

Furthermore, it is not just energy price volatility and rising CO_2 prices that could weigh on the shoulders of poorer households in future: the extension of the European ETS could too. A key feature of the 'Fit for 55' package is the extension of the EU ETS to new sectors. Whereas the ETS currently covers the power and heavy industry sectors, accounting for 40 per cent of emissions, the European Green Deal proposes to extend the ETS also to the aviation, maritime, building and transport sectors, adding a further 50 per cent of emissions within the remit of the European cap-and-trade system (Simon 2021). In particular, an additional ETS to cover buildings and transport is proposed to enter into force in 2025, when suppliers would start buying emission permits.[10] This extra cost could end up being offloaded by companies onto households, thus aggravating energy poverty and mobility inequalities. The same is true also of the phasing out of free allowances for industries in sectors already covered by the ETS. In fairness, the Commission's 'Fit for 55' proposal implicitly caters for this, foreseeing a separate ETS for buildings and transport, featuring a lower carbon price to avoid a one-shot tax on these sectors from zero to €100 – or more – per CO_2 ton. Yet even this is raising eyebrows, in the context of surging energy prices, with opposition mounting, notably in France and Spain, as well as in the European Parliament, in which several political groupings have long been concerned with the distributional effects of an extended ETS. French member of the European Parliament Pascal Canfin did not mince his words, defining the extension of ETS to the transport and building sectors as 'political suicide' (Simon 2021).

If carbon prices do rise and are extended to other sectors to accelerate the transition, yet they risk having regressive socio-economic effects, the political repercussions are not hard to predict. The *gilets jaunes* protest in France in 2019, in which around 300,000 protesters wearing fluorescent yellow vests brought the government to its knees by occupying round-abouts, blocking roads and shopping centres over the course of several months, serves as a stark reminder. The root causes of the protest movement are multifaceted and complex, but the trigger was French President Emmanuel Macron's increase in fuel taxes in a publicly asserted effort to finance the national ecological transition (Public Senat 2019). Over time, the movement morphed into an anti-systemic one, peppered with conspiracy theorists, far-right white supremacists and far-left nationalists that mobilized largely through social media. Its heterogeneity (Institut Montaigne 2019) made it hard to grapple with, engage or appease. This said, the movement did channel the anger of economically vulnerable low- and middle-income segments, notably those in rural areas, whose mobility hinges on private cars in view of degraded public transport services (La Fabrique de la Cité 2020). The juxta-position made by those on the streets was that whereas the elites could afford to worry about the end of the world, they were all consumed by their struggle to reach the end of the month. The energy transition is something that they would have liked ideally, but they simply could not afford.

Back then, the *gilets jaunes* did not channel their anger against the EU. Their prime target was President Macron and the liberalism that he embodies. However, it is only afterwards that the EU elevated the green transition into its dominant narrative. A *gilets jaunes* protest in France or elsewhere today could most likely have a more pronounced Eurosceptic twist.

Climate, energy transition and Euroscepticism

Climate and the energy transition constitute good material for nationalist populist Eurosceptic parties. Climate has been on the nationalist populist radar for some time. Several such parties have challenged the scientific consensus on climate change (Schaller and Carius 2019). Their scepticism has not been uniform, however, with the German Alternative für Deutschland (AfD) or the Dutch Partij voor de Vrijheid questioning the actuality of anthropogenic warming, and the French National Rally or the Spanish Vox ignoring the transnational nature of the crisis and proposing national(ist) responses to it.

Moreover, these parties have generally shunned multilateral climate solutions, focusing on preserving the national environmental heritage or reducing international interdependence. In the case of the Italian Lega, this has meant scepticism about renewables, which rely on foreign (Chinese) technologies to the detriment of Italian business. Renewables have been opposed by the Danish Freedom Party, the Swedish Democrats and, more recently, the French National Rally, because of their impact on the 'traditional landscape'. In France, opposition to renewables has focused in particular on wind energy, with Éric Zemmour, a TV show polemicist and 2022 presidential candidate, defining it as a 'catastrophe' that would destroy the French coastline (White and Mallet 2021).

Whereas energy and climate populism has typically tilted towards climate scepticism and opposition to decarbonization, some nationalist populist parties have embraced the energy transition, and particularly the growth of renewables, positing equally ideological arguments. The True Finns, for instance, strongly support renewables because they help reduce interdependence with the 'Islamic' Gulf (Schaller and Carius 2019). More broadly, the move towards renewables produced mainly within countries rather than traded between them is looked at favourably by parties supporting trade protectionism and closure. As discussed at length in chapter

4, the more decentralized nature of a green energy system is backed by nationalist parties in light of their broader support for protectionism and sovereigntist political agendas. Other parties, on the populist left have echoed similar approaches. Prior to its governmental turn, the Italian Five Star Movement advocated various forms of 'happy de-growth', favouring reduced and more localized production and consumption patterns. Support for renewables meshed with the drive for de-globalization and protectionism. With nationalist populist forces occupying both extremes of the climate and energy transition divide, the space for political dialogue, facts and truth has narrowed. This has started giving rise to a green wedge (Fieschi 2022), in which ideological arguments linking climate and social justice have been used by both populist proponents and opponents of climate action.

At the same time, as the consensus around climate change grew in Europe, those nationalist populists who toyed with climate denialist views, toned these down, as they no longer paid off electorally. Rather than refuting anthropogenic climate change, they began lashing out against the establishment's responses to it. With a few exceptions like the German AfD, their case has differed from the traditional right-wing argument in the US, whose climate scepticism has focused on the concern that climate policies, revolving around regulation and the role of government, would stifle the market and therefore growth. The European populist opposition to climate action has mainly highlighted its adverse consequences for disadvantaged industries, regions and groups (Żuk and Szulecki 2020). In this crucial respect, their critique is deeply embedded in the broader nationalist populist agenda that has accused European policies for their injustice, lack of accountability and detachment from the real interests of the 'people'.

This is where the nexus linking the European energy transition and its socio-economic effects kicks in. Much like Eurosceptic nationalists have criticized the establishment's approaches to the European economic and migration crises, they are now making similar points about the energy transition. These range from structural arguments opposing the potentially regressive effects of climate policies to ideological critiques of the cosmopolitan elite's climate agenda, backed by its 'expert' acolytes (Lockwood 2018). Both the structural and ideological lines of argument focus on the establishment's alleged offloading of the transition's burden onto the shoulders of the left behind. Notably carbon pricing and taxation policies – which represent key elements of the European transition policy mix – are portrayed as establishment attacks on the impoverished lower and middle classes (Serhan 2021). The fact that the EU champions such climate measures comes as a welcome bonus. By picking a fight with the European Green Deal because it allegedly impinges on national sovereignty and has regressive effects, Eurosceptics seek to kill two birds with one stone, both the energy transition and the establishment, bureaucratic and liberal EU itself.

Indeed, the first signs are there to be seen. When the European Green Deal was first presented, the French National Rally (2020) lambasted its globalist ideals, used as a pretext for the EU's never-ending power grab. To the French far left, Jean-Luc Mélenchon (2019) defined the EU's green agenda as a 'liberal hallucination', claiming that to embark on positive change what was necessary was not 'more Europe', but more state, sovereignty and solidarity. Authoritarian leaders in the EU, like Hungarian Prime Minister Viktor Orbán, criticized the European Green Deal, describing it as a 'foolish plan' in which 'everyone would be put at a disadvantage' (Hungary Today 2021). His views were echoed by then nationalist Czech Prime Minister Andrej Babiš (2021a), who claimed that the EU's green agenda was 'not a deal, but an ideology' that would result in a 'European Green Suicide', reassuring Czech citizens that 'we will not let you become poorer because of environmental madness' (Babiš 2021b). As of today, these are only a few scattered signals of how Eurosceptic parties and leaders may use their rejection of Europe's energy transition as their new rallying cry. But as the European energy transition translates from rhetoric to reality, a more adamant Eurosceptic opposition, as well as a more vocal political dialectic and probably greater divisions between Member States and within institutions, can be expected.

Addressing the socio-economic costs and political consequences of the transition

Aware of the socio-economic and the ensuing political risks of the 2021–2 energy price spike, European governments and the Commission responded. Member States like Italy and Sweden subsidized consumer bills of the most disadvantaged households and energy-intensive industries. Others like France and Spain set gas price caps for the first time, with the former also ramping up the production of nuclear-generated electricity. All in all, these ad hoc measures cost tens of billions of euros over the course of one autumn/winter, even before Russia's war in Ukraine erupted.

Were this crisis simply the product of a temporary mismatch between demand and supply, which energy markets have often experienced in the past, temporary national measures to subsidize the worst off would be enough. Yet if the profound volatility in prices is to become a structural feature of the transition, ad hoc fiscal measures will be insufficient at best, paradoxical at worst. To the extent that the energy transition will be characterized by institutions and investors turning away from fossil fuels whilst low-carbon sources will have not filled the supply gap in the short term, volatility will follow. Price spikes and collapses are to be expected as a consequence of the cyclical ups and downs of demand. This may be the first of many price spikes, which could end up weighing disproportionally on vulnerable territories, industries and communities, making the ad hoc fiscal measures taken by individual Member States woefully inadequate. Ex post ad hoc fiscal measures to alleviate the costs of utility bills for households and businesses would be inadequate both quantitatively and qualitatively, given they imply an indirect subsidy to the consumption of fossil fuels while Europe seeks to transition away from them.

The risk that the transition could have structurally regressive socio-economic effects and that these might be exploited by Eurosceptic forces is on the EU's radar. Support measures aimed at contrasting the potentially regressive effects of the transition represent an important feature of the European Green Deal, and part of the revenues from the EU's carbon pricing policies are foreseen precisely as carbon dividends to support those most in need. In the 2021–2 energy crisis, the Commission endorsed the short-term measures taken by European governments to contain the effects of the price rise and gave reassurance that these would not be subject to state aid rules and could be covered partly by the higher than expected revenues generated from the ETS auctions (European Commission 2021b).

In future, part of the revenues generated by the expanded ETS and the CBAM could be channelled to low-income territories and individuals. As part of its 'own resources' package, the Commission aims to channel €12 billion per year from the ETS's revenues to disadvantaged groups from 2027 onwards, adding further billions of euros when CBAM eventually becomes operational that year. The notion of carbon dividends will no doubt meet obstacles along the way. Within the EU, for instance, as decarbonization proceeds, the revenues from ETS auctions should fall too, reducing the dividends used for redistributive purposes. Externally, using CBAM revenues for redistribution may complicate the path to making the mechanism compliant with the rules of the World Trade Organization (WTO) that would rule against CBAM if this were considered protectionist and discriminatory (Sapir 2021). More broadly, there is deep political disagreement in the EU regarding how the potentially regressive effects of the transition should be redressed. Some Member States, including Poland and France, believe these should be tackled by EU energy and climate policies themselves. Others, notably Germany and more generally Member States with greater fiscal space to manoeuvre, believe that other national instruments should come to the rescue. These and other challenges would need to be overcome in the years ahead. What is clear, however, is that significant public funds will be necessary to counter the potentially adverse socio-economic effects of the transition.

For this purpose, the Commission has devised several instruments. The Just Transition Fund aims at supporting carbon-intensive territories in their decarbonization journey, both by easing the phasing out of coal, peat and shale, and by decarbonizing hard-to-abate industries and/or replacing them with low-carbon jobs and activities. Member States would be eligible for these funds upon submission of decommissioning and socio-economic adaptation plans. The Just Transition Fund is meant to prevent the exacerbation of territorial disparities within Member States, but also, crucially, between them, notably between some heavy coal consumers in eastern Europe and west European countries that are better equipped to face the transition (Eyl-Mazzega and Mathieu 2020). Poland, for instance, with its over 115,000 coal workers, is expected to be the main beneficiary from the Fund (Euractiv 2021). Given that access to 50 per cent of the money is conditional on approval of the 2050 climate neutrality goal, the Transition Fund could also embed net neutrality in those Member States – like Poland – that are yet to commit to it formally. The Transition Fund would work alongside the modernization fund, expected to reach €14 billion over this decade to support lower-income Member States in eastern Europe to modernize their energy grids, invest in energy efficiency and storage, as well as redeploy and reskill workers in carbon-dependent regions (European Commission 2021d).

The political and geopolitical relevance of these funding instruments is key to the future of Europe: the worry is that without them the existing cleavage between eastern and western Europe that emerged with the migration crisis and the democratic backsliding in Poland and Hungary would deepen in light of Europe's energy transition. As the Union seeks to consolidate its newfound sense of solidarity against the backdrop of the pandemic and the war in Ukraine, the last thing it needs is to sow the seeds of division again by cementing a climate cleavage between east and west.

As discussed in chapter 1, the EU's internal unity on climate policy is also a precondition for its effective external leadership. The EU has successfully led by example in the world when Member States have stood united. Its global leadership weakened when its internal consensus became more fragile, notably in the early post-eastern enlargement years as highlighted by the debacle at COP15 in Copenhagen. In the run-up to and since the 2015 Paris Agreement, the EU's internal unity was strengthened once again, and galvanized to promote an ambitious global agenda. Fostering such unity remains essential both for an internal revival of the European project coalescing around the green agenda, and for its climate leadership in the world.

Addressing the potentially regressive effects of the energy transition requires both a focus on territory as well as on people. The Climate Social Fund is intended to support those individuals in the EU who risk being disproportionally hit by the transition, and particularly by the extension of carbon pricing to buildings and transport. This would include vulnerable citizens, as well as small enterprises and transport users. The fund would thus assist the adaptation of lower-income citizens to low-carbon livelihoods, including switching to green transport, improving the energy efficiency of private housing as well as income support when necessary. Exactly how the Fund will be targeted and channelled to individual European citizens remains to be seen, as well as the precise balance between EU funds and complementary ones from Member States. However, the Climate Social Fund with its focus on people, alongside the Just Transition Fund with its emphasis on territory, represent indispensable socio-economic – and political – components of the European Green Deal.

Policy and political avenues ahead

All this goes in the right direction. These are all measures that contribute to making the green transition socially and politically acceptable, which in turn is a prerequisite for making this policy-driven, and not just technology-driven, change possible. The question is whether these measures are sufficient (Bianchi and Colantoni 2021). The Just Transition Fund was eventually cut to €17.5 billion by the EU Council of Ministers, from the Commission's proposal of €40 billion (European Parliament 2021). Some Member States, notably the so-called 'frugals' from northern Europe, always eye with suspicion any form of financial transfer from richer to poorer European countries.

This connects to the longer-term and more structural question of the EU's future fiscal rules, which could lead to a revived north–south divide within the Eurozone. Against the backdrop of the pandemic, the EU both suspended its Stability and Growth Pact rules that foresee strict limits to public deficits and debt levels and approved its significant NextGenerationEU Fund, a large chunk of which is to be channelled to decarbonization. This went far in healing the divisions that had split the Eurozone during the sovereign debt crisis, especially between northern and southern Member States. However, NextGenerationEU funds will not last forever and the Eurozone is yet to agree on a new set of fiscal rules. Were the 'old' pre-pandemic deficit and debt rules to be reapplied, the north–south cleavage over fiscal policy would risk becoming green. This is because only those Member States with greater fiscal space would be able to continue investing in decarbonization after the EU's funds run out. It is in this spirit that France and Italy proposed a revision to the Stability and Growth Pact, inserting a golden rule that would exempt green investments from the deficit calculus. Only if EU institutions are entrusted with own resources to provide ongoing support for a just transition and Member States are provided with equal fiscal opportunities to invest in it, can the EU remain united throughout this tumultuous journey in the decades ahead.

Alongside the overall amount of transition funds available, the question is how these will be spent. Whereas the energy transition will see the phasing out of some jobs and industries and the creation of new ones, with the Just Transition Fund intending to alleviate that journey, it does not necessarily mean that industries and jobs will be recreated in the same places. As Poland, for instance, gradually moves away from coal and invests in the development of its promising wind industry, the regions that will be affected by these industrial phasing outs and ins are not always the same. Likewise, whereas the net loss of jobs in the switch to the manufacturing of fossil-fuelled versus electric vehicles can be more than compensated by new jobs in the battery, storage or recycling industries, the workers involved will not always be the same. In other words, whereas the Just Transition Fund may play an important role in supporting the energy transition of certain countries, notably in eastern Europe, it does not automatically mean that it can avoid growing disparities within them and the ensuing grievances these can generate.

When it comes to support for individuals in Member States through the Climate Social Fund, the EU has envisaged €59 billion, reaching 144 billion if Member State contributions are included. While significant at face value, much more may be necessary in the years ahead. In just the autumn/ winter of one year – and before the Ukraine war began – one Member State – Italy – spent over €10 billion to subsidize households given the spike in energy prices. If energy price spikes become more frequent and protracted, and the cost of carbon steadily rises and will need to be compensated to avoid weighing disproportionally on the shoulders of lower-income individuals, €144 billion over seven years for twenty-seven states is likely to be insufficient. All the more so given the relatively small role played by carbon pricing in the 2021–2 crisis, a role that is only expected to grow in future as CO_2 prices continue to increase. In light of the fact that the increase in power prices in the 2021–2 crisis is only marginally tied to the increase in carbon prices, but that the costs of fossil-sourced power and transport will increase in future because of the lower caps and extended range of the ETS, the funds necessary to redress the resulting socio-economic disparities would need to be significantly higher.

Furthermore, all this does not take into account the adverse effects of the climate crisis itself, notably how the EU will assist disadvantaged territories and individuals meet the costs of climate adaptation. Sea levels are expected to rise almost everywhere in Europe, forest fires are ever more frequent in the Mediterranean, while the floods in Germany and Belgium in the summer of 2021 are only a first dramatic reminder of what lies in store. The German insurance firm Munich Re calculated that natural disasters in 2021 cost €247 billion globally and that Germany's summer floods caused €35 billion worth of damage (DW 2021). This suggests that the tailored financial instruments the EU is developing are important, but only represent the tip of the iceberg of the funding that Europe will need to cough up to make the transition socio-economically, and thus politically, sustainable. What will be necessary is not just the development of specific and inevitably limited European-level financial instruments financed by the EU's own resources, but also the 'greening' of all existing funding instruments at the Union's disposal. The earmarking of 30 per cent of the far more significant 2021–7 €330 billion cohesion funds to the goal of decarbonization is a model in this respect. In other words, both EU and Member State public finances will need to mainstream climate and the transition to equip themselves for the revolutionary change in store.

Financial compensation as well as non-fiscal measures such as education and training to redress the potential disparities generated by the transition are essential. Alongside these, what more could be done to hedge against the endemic volatility in commodity prices that is likely to accompany the energy transition? A much-discussed avenue revolves around European strategic gas reserves. The debate over European gas reserves is a long-standing one, traditionally pushed by some Member States either because of their non-reliance on gas for electricity generation – France – or because of insufficient national storage capacities – Spain, Greece, Romania and the Czech Republic. The idea makes intuitive sense. In an integrated single market, ensuring that at least strategic reserves are jointly procured by and shared between Member States would increase the EU's bargaining power vis-à-vis producing states, as well as mitigate against volatility in times of supply disruptions, major infrastructure outages or particularly cold spells. Unsurprisingly, the 2021–2 energy price crisis and the ensuing Russian invasion of Ukraine reawakened and spearheaded this long-standing debate, with policy avenues headed towards joint gas storage and possibly even procurement. Moving in this direction requires overcoming deep political, infrastructural and institutional hurdles. However, doing so is essential if the EU is to become energy independent.

Another possible route to reduce volatility is the rethinking of long-term relationships with supplier countries. Turning again to the 2021–2 energy supply crunch, Russian President Putin sneered at the Union and its progressive shift towards spot gas markets and away from long-term contracts in recent years (Gotev 2021). It is true that the race for gas on spot markets in Europe and Asia contributed to the price spike in the autumn of 2021. However, it would be paradoxical, to say the least, if the Union were to simply incentivize long-term contracts as we know them today. Tying oneself to decades-long gas contracts as the EU seeks to decarbonize would make little sense. The European Commission in fact is moving in the opposite direction, going as far as proposing a ban on long-term contracts for unabated gas after 2049. Whether taking the form of a ban or other measures, the question indeed is that of fundamentally rethinking such contracts. The EU could continue to support long-term relationships that help shield against growing volatility, while avoiding getting locked into a long-term fossil future. Rethought long-term relationships could feature a progressively growing share of low-carbon supplies – including gas with CCS and renewable sources – over time, as well as technological transfers, normative-regulatory exchange and cooperation on energy efficiency and the interconnectivity of electricity grids (Pastukhova et al. 2020). This could both contribute to a containment of price volatility, protecting the more vulnerable groups in the Union, while incentivizing the energy transition beyond its borders, a topic explored in detail in chapter 3.

Alongside facts, funds and contracts, what is equally important is European storytelling (Schaller and Carius 2019). It is striking how the likes of President Putin have attributed the 2021–2 energy price rise also to the transition, and particularly to the growing reliance on renewables in Europe. In the words of the Russian president: 'due to a decrease in wind farm generation, there was a shortage of electricity on the European market. Prices soared, which triggered a spike in natural gas prices on the spot market' (Gotev 2021). There is a connection between a lower than expected generation of wind energy in the summer of 2021 and the price spike in autumn as discussed earlier. But, as explained above, it is part of a far more complex story, which cannot be reduced to the fact that the price rise – and its adverse distributional effects – are due to the European Green Deal. EU institutions have pushed back, with Energy Commissioner Kadri Simson arguing that this crisis, and its regressive effects, warrant a speeding up, not a slowing down of the transition (Thebault 2021). The cost, the argument goes, is driven by insufficient change, not the opposite. Communicating complexity is never easy, however, and it requires a concerted effort by international and European institutions, business, the media and civil society. It entails confronting openly the policy trade-offs to be made as much as countering fake news both on climate change and on the energy transition, as well as explaining the positive potential underpinning change.

This means that while a narrative on the brave new green world is necessary, alone, it is insufficient. The Union so far has developed a powerful narrative for a green Europe, which, as discussed in chapter 1, lies at the heart of a revival of the European project. This is necessary but insufficient. If change is to happen, the EU will need to develop a compelling narrative not just for a 'green Europe', but for a 'transition Europe' too. Due to its complexity and embedded trade-offs, this is going to be a harder story to sell. However, it is just as, if not more, important given the decades-long journey ahead.

Communicating defensively – i.e., defending the transition – is important. Just as important is communicating offensively, by highlighting proactively the economic and healthcare costs and regressive effects of inaction. On the one hand, whereas a policy-driven transition seeks to be faster than what it would otherwise be if left to market forces alone, the economy and technology would lead to change anyhow over a longer period of time. In other words, the coal miners who risk losing their jobs in the coming years would run that risk regardless of the EU's energy and climate policies. The latter certainly means frontloading that risk: those jobs will be lost earlier. However, it also puts those socio-economic risks into the political spotlight, leading to policy measures aimed at mitigating and compensating for those losses. In other words, it is thanks to the fact that this energy transition is a political and policy-driven process and not only a technology- and market-driven one that its socio-economic disruptions are more likely to be discussed and addressed.

On the other hand, not acting to address climate change would have even more damaging and regressive effects. It is the same vulnerable individuals and households who will suffer more from the lack of transition, and therefore from an exacerbation of the climate crisis. They will be the ones more likely to be displaced by rising sea levels, whose uninsured properties will be destroyed by extreme weather events, and who will lose their livelihoods because of the changing climate (Zachmann et al. 2018). Pre-existing inequality means that disadvantaged groups are and will suffer disproportionately more from climate change, exacerbating disparities. So far, the political narrative of the costs of remaining 'brown' has failed to gain political traction, while that of the cost of becoming 'green', particularly if it is the EU doing so, is increasingly being voiced by the likes of Putin, Le Pen or Orbán. Succeeding in this regard is key if the energy transition is to become socially, economically and therefore politically sustainable, and thus to actually happen, given the policy-driven nature of the process.

Much like the EU has developed its strategic communication activities to counter Russia's disinformation campaigns concerning migration, Catalunya or Brexit, China's propaganda in the context of Covid-19, not to mention the disinformation emanating from Da'esh and other terrorist groups, the same will need to be done about the energy transition and climate. Strategic communication activities in Europe aimed both at countering disinformation and engaging in positive storytelling should increasingly factor the energy transition into their work.

* * * * * * *

Liberal democracy and the European Union have been under attack by authoritarian countries and by nationalist populists, who, in Europe, have invariably been Eurosceptic. Their narrative has focused on the failures of liberal democratic systems and policies, first and foremost of the EU itself. While manipulating, distorting and exaggerating facts, these political forces have often leveraged real fragilities and failures. In particular, the EU's handling of the sovereign debt crisis and of the migration crisis laid bare socio-economic disparities, awakened identity fears and revealed a stark lack of solidarity between Member States. The EU, and particularly its institutions, appeared ever more distant and detached from the lives and daily problems of European citizens.

Solidarity amongst Europeans was rediscovered over the course of the pandemic. Driven by an awareness that Europeans could navigate Covid-19 only by standing united, the joint European health and above all economic response to the crisis lifted the EU from its knees. The Ukraine war welded Europeans even closer together. The EU's green narrative can provide it with the opportunity to fly once again.

However, the green revival of the European project is not a done deal. The energy transition aimed at achieving net zero greenhouse gas emissions by 2050 implies a true revolution on the European continent and beyond. Like in all revolutions, there will be winners and losers. The losers from this transition, as is often the case, could be the most vulnerable groups in society. If left unaddressed, these disadvantaged groups will make their voices heard. And Eurosceptic forces, which are on the lookout for a new political banner to galvanize public support, already begin to see in the energy transition an opportunity to attack both the green agenda and the EU at its helm.

The socio-economic challenges and the ensuing political risks created by the transition are possible but not inevitable. If the EU takes the necessary steps to proactively mitigate the structural energy price volatility that lies ahead, adopts significant compensatory measures to support vulnerable territories and groups, and develops a compelling narrative not just for a bright green future but for the bumpy yet necessary transition ahead, the promise of a European green revival can become a reality.

3

A Green Europe in a Troubled Neighbourhood

Europe's energy transition could have internal socio-economic and ensuing political consequences, which drive at the heart of liberal democracy and European integration. These internal implications must be addressed promptly and proactively if both are to be strengthened. However, this political and policy-driven transition will also have profound international implications, starting with Europe's surrounding regions, both to the east and south. Addressing these regional consequences will be equally important if the EU's energy transition is to be geopolitically sustainable and thus politically sustained by institutions and societies in the decades ahead. This will require placing far more emphasis on climate and energy in European foreign policy and developing a joined-up approach between the European climate and energy policies and all other dimensions of the EU's external action.

Europe's neighbours and the EU's relations with them have been troubled for some time. From the fragility of states and societies in eastern Europe, the Caucasus, north and sub-Saharan Africa and the Middle East, to the persistence, outburst or re-eruption of violent conflict within and between states, to the troubled relationship with Turkey, through to the open confrontation with Russia, the Union is surrounded by a proverbial ring of fire (*The Economist* 2014). The EU's energy transition interlocks with these regional dynamics, at times exacerbating them while also offering new avenues for resilience and cooperation.

Given the EU's level of climate ambition, leadership as well as its weight as a major global economy and energy consumer, the EU's transition will impact others, especially its eastern and southern neighbours. It can do so through two main channels of influence. Viewed through a material lens, the EU's transition and the changing patterns of European energy consumption that come with it, will influence profoundly global energy markets. If the EU's consumption of fossil sources falls, and that of low-carbon alternatives grows to the levels envisaged by the European Green Deal, these changing demand patterns will reverberate across global energy markets. More specifically, they will affect different geographies, particularly those that are more dependent on or interdependent with the EU. These effects relate both to the energy transitions of these countries, providing EU leverage on them, as well as their broader state and societal strength and fragility. In view of the EU's overall weight in global energy markets, as well as its relative material influence on its surrounding regions, its energy transition will shape profoundly the trajectory of the global energy system as well as the development and governance trajectories of its neighbours near and far.

Viewed through an ideational lens, the EU's energy transition will act as a trailblazer and a potential model for others to follow. As of today, over three quarters of the world have committed to climate neutrality, covering over 90 per cent of global emissions, with the number of countries making pledges constantly growing. However, as discussed in chapter 1, the European Green Deal represents the first attempt worldwide to go beyond pledges, outlining a concrete plan with specific targets and proposed policies in line with the Paris Agreement. The EU's plan is certainly not the only model, and it is already clear that other countries will follow different routes, based on their different capacities, political cultures and comparative advantages.

However, insofar as the EU has set the first concrete path to reach net neutrality, and given the deep and long-standing interdependence between the EU and its neighbours and the former's soft power over them, the European Green Deal will inevitably influence the EU's surrounding regions.

Moreover, this influence will go well beyond the energy transition itself, extending to the broader development and political trajectories of countries beyond the EU's borders. This is because this energy transformation implies not just the switch to different energy sources. With the significant uptick in electrification, it entails a transformation of energy systems – from grids, distribution and storage capacities to the governance necessary to operate these. By affecting the energy transitions beyond its borders, the European Green Deal has the potential to influence deeply the overall development paths and governance systems of the EU's troubled neighbours.

To an extent, this is recognized. EU policies towards the Western Balkans, Turkey, eastern Europe, the Caucasus, the southern Mediterranean and sub-Saharan Africa all feature climate and energy in their mix. From the focus on regulatory harmonization, infrastructure development and energy efficiency, to the support for renewables and climate adaptation, the EU is increasingly weaving the transition into its foreign policy mix. EU climate and energy initiatives are more pronounced in some regions than in others, reflecting the EU's geographic priorities. The most ambitious and concrete EU endeavours concentrate on the Western Balkans, which for the 2020–7 period foresee a €9 billion Economic and Investment Plan and a tailored green agenda, followed by eastern Europe, select countries in the southern Mediterranean and Africa (Teevan et al. 2021). While a differentiated geographic attention is inevitable and country-specific approaches are desirable, the scale and mainstreaming of the EU's climate and energy components within EU foreign policy will need to be significantly increased.

Thinking through the multifaceted implications of Europe's transition in its troubled neighbourhood helps to unpack what these policy components could look like. Following this line of thought, this chapter examines the different dimensions linking Europe's decarbonization and its complex relations with neighbouring countries. On the one hand, the EU's internal energy transition affects profoundly its neighbours, both as a direct consequence of European climate and energy policies and as an indirect effect of the Union's overall weight in global energy markets. Of particular note are those neighbours with carbon-intensive relationships with the EU. Amongst these are fossil fuel producing states that already feature economic, social, governance and security fragilities, states with a track record of weaponizing energy, as well as states that will be hit most severely by the EU's CBAM. Also important, however, are countries that, while poor in fossil fuels, have significant potential in a decarbonized world. It is essential to factor into a green Europe these regional challenges and opportunities. On the other hand, the climate crisis is already felt in the EU's surrounding regions, notably in the south. This calls for a strong focus of European foreign policy on climate action, with climate adaptation becoming part of the core rather than an addendum to the EU's joined-up external action (EEAS 2016). As this chapter explains, all this requires developing a far more pronounced external dimension of the European Green Deal than what is currently envisaged, as well as its integration in all dimensions of European foreign policy.

Supporting the transformation of energy and governance systems beyond EU borders

As the EU decarbonizes internally, it will need to develop equally pronounced external climate mitigation policies, especially in its neighbourhood. This implies elevating the energy transition of neighbouring regions as a priority of EU external action. However, given the nature of the transition, this will require a comprehensive push to reform the energy and governance systems of these countries.

The International Renewable Energy Agency (IRENA 2019a) that conducted a first study addressing the geopolitics of renewable energy sources, while upbeat, did not gloss over the complexities of the transition. It acknowledged, for instance, that the higher technological and digital content of a greener world would come with new fragilities, from cyber, electric grid security and privacy, to the need to access critical minerals notably for the development of battery storage and clean energy technologies more broadly. I shall return to this in chapter 4. The IRENA study also recognized that countries with a technological edge would wield greater power in a green world, risking an exacerbation of disparities between developed and developing economies. In short, the countries that will be better equipped to face the transition and that will thrive in a decarbonized world are those with stronger economies, more resilient infrastructures and more effective governance systems. This means that achieving decarbonization does not imply only the replacement of fossil fuels with decarbonized energy sources but rather the deep transformation of energy systems, with all the economic, infrastructural and governance changes entailed.

Bearing this in mind, if the EU is to promote climate mitigation beyond its borders, its support for third countries must both foresee significant climate and energy transition funds as well as extend well beyond the energy sector, permeating across governance, security, infrastructure, economy and technology too (Franza et al. 2020). Precisely because the current transition is a systemic one, the EU's green partnerships must cover the entire governance systems of third countries. This requires targeted financial and technical EU support for reforming the energy, environmental, infrastructural and institutional systems of neighbouring countries, as well as the integration of these objectives within European security, development, digital and trade policies.

Two examples highlight the approach that is necessary as well as the difficult trade-offs involved. The links between the energy transition, digital and security policies are clear, whereby the greening of supply chains and the management of demand are possible only through a digitalized energy system, especially a digitalized distribution system (Tagliapietra et al. 2019). A decarbonized and digitalized energy system is also one that requires stepped-up cyber security policies to protect critical infrastructure, with new fragilities emerging as a result of decentralization and widespread electrification (Morningstar et al. 2020). In most countries, electricity grids are at far greater risk of cyberattacks than hydrocarbons. This is understood within the EU, in which the European Green Deal and the digital agenda are seen as being closely intertwined. However, the same applies internationally as well, as European foreign policy seeks to promote decarbonization in its surrounding regions. Yet in the EU's external action, much remains to be done to connect the dots. Not only does the EU pursue very few climate mitigation projects beyond its borders, but rarely, if ever, do these feature digital governance.

The nexus between energy, trade, development and diplomacy is another case in point (Grübler et al. 2021). There is a general recognition of the need to 'green' the EU's trade agreements. Currently EU trade agreements include sustainable development chapters, which seek to facilitate trade in climate-friendly goods. These have not been particularly effective to date. The aim now is to strengthen the climate anchoring of EU trade by including alignment with the Paris Agreement as an essential element of future free trade agreements (FTAs). But much like the human rights articles that are essential in all EU FTAs, probably there would be weak enforcement mechanisms in these climate clauses. Specifically, it is very difficult – if not outright impossible – to suspend the application of these agreements in the event of non-compliance, with environmental provisions being exempt from dispute settlement mechanisms. Moreover, there is an unspoken tension between expanding trade – which still typically comes with increased greenhouse gas emissions – and decarbonization.

This said, addressing non-compliance with climate expectations can take different forms. The withdrawal of benefits – such as the suspension of a free trade agreement – or the infliction of costs – such as sanctions – is one way forward. In a few cases in the neighbourhood, such as Russia or

Belarus, the EU opts for this coercive route. However, in most other cases, the EU shies away from suspending cooperation agreements or imposing economic sanctions on third countries, notwithstanding gross violations of human rights and democracy all around. With trade policy being primarily concerned with trade itself rather than being viewed as an instrument to promote human rights, democracy or sustainability, the EU often resists leveraging trade in the pursuit of other foreign policy objectives.

In most cases, the EU highlights its preference for dialogue and cooperation, revolving around diplomacy and development. This is not the place to elaborate on the relative effectiveness of sticks versus carrots in European foreign policy. There is an established literature on this subject (Lavenex 2004; Nye 2004; Lucarelli and Manners 2006; Tocci 2007). Suffice it to say here that, effectiveness aside, the exercise of European soft power hinges on the deployment of significant diplomatic and development resources to influence governments, strengthen governance and support civil society. In other words, soft power requires a strong diplomatic and development presence and push. Applied to the case of sustainability and decarbonization, if the EU is unable or unwilling to suspend cooperation or impose sanctions in response to insufficient work on climate mitigation, it must put the green agenda at the core of its diplomatic engagement and development cooperation with third countries. This is far from being true today.

Working in this direction is key if the sustainable development goals are to be pursued in all their dimensions. In particular, if the world is to embark on decarbonization, but also ensure energy access to all, the EU's support for its neighbours cannot stop at the promotion of clean energy sources but requires contributing to a radical rethink of governance systems using all external action tools at the EU's disposal. This is because of the inadequacy of the current system not just to provide clean energy, but to ensure sufficient energy *tout court*. In other words, rethinking energy systems is crucial both from the perspective of the energy transition and of energy access. Framing EU support to this effect through this double lens – energy transition *and* access – is essential especially in a continent like Africa, where the latter weighs far more than the former as a public policy concern (Pistelli 2020). With two thirds of the population with no access to electricity, 90 per cent with no access to clean cooking (IEA 2018), and the pandemic having slowed down considerably electrification, work on energy access and energy transition can only go hand in hand. The challenge ahead is huge and can only be addressed if the EU prioritizes access to clean energy beyond its borders through the deployment of all its external action instruments.

Within this overarching foreign policy framework, EU external action needs to distinguish between five different subsets of countries and issues.

Addressing fragilities in neighbouring regions

First, as the EU proceeds with its energy transition, it must specifically cater for the vulnerabilities that it might inadvertently create and exacerbate within certain fossil fuel producing countries. Like all revolutionary changes, the energy transition will not be a gala dinner for all. Much like the energy transition risks having adverse socio-economic distributional effects within the EU, the same argument applies beyond the EU's borders. Addressing the external fragilities that will be indirectly created or exacerbated by the EU's energy transition will be just as important as intervening on the internal ones.

A critical and much discussed category of countries is the petrostates, many of which surround Europe and some of which are highly dependent on the EU for their energy exports and public finances in general. This is the category within which the most acute fragilities are expected to emerge, not least because of the pre-existing state and societal vulnerabilities in many of these countries. Over the decades, the concentration of oil and gas resources in certain countries led to significant socio-economic development, but also to profound economic and governance fragilities. Some of these rentier states may even have suffered from variants of the Dutch disease (Oomes and Kalcheva 2007; Gasmi and Laourari 2017).

However, such fragilities are not equally spread. As Goldthau and Westphal (2019) persuasively argue, the energy transition will affect existing petrostates in fundamentally different ways. In some countries, the transition may provide fossil fuel industries with a temporary boost. This is true of petrostates that have developed refining capacities, particularly as these are being squeezed out of Europe by falling oil demand, volatile commodity prices and higher carbon prices. The Gulf countries stand out in this respect. More broadly, some Gulf states have outlined economic diversification strategies, with countries like the United Arab Emirates also beginning to put them into practice. Temporarily shielded from the turbulence in store also applies to gas-rich countries like Qatar. This is because European gas demand is expected to decrease only after 2030, according to the European Green Deal's own targets in view of the ongoing coal-to-gas switch and the nuclear phase-outs in several countries. Europe's turn away from Russian gas raises the value of Qatar's gas further. In a different way, this is also true of Azerbaijan. Whereas Azerbaijan does not enjoy the low extraction costs of countries like Qatar, and as a result is struggling to gain market share,[11] Baku sits comfortably on 25-year long-term energy contracts, providing it ample time and space to adapt. More generally, fossil fuel countries with low break-even prices, rich sovereign wealth funds and developed national oil companies can retain and even capture market share in the years ahead. As price volatility endures, driving out of the market countries with more expensive sources, and as international energy companies cut back their upstream investments that increasingly struggle to access capital from the market, some fossil fuel countries may even temporarily benefit from the energy transition. Over the next decade, many of these countries will face the 'green paradox' (Sinn 2012). Precisely in anticipation of the decline in demand and ensuing revenues that is expected to come with the transition, these producer countries will be induced – and will be able to afford – to increase their fossil fuel production and their savings.

This said, most fossil fuel producing states are in for a rough ride much earlier. With few exceptions – like the UAE, which has successfully begun diversifying, or even Iran, which while economically weak has been forced to diversify under the weight of economic sanctions – most petrostates have hardly begun their transition journeys. Price volatility in energy markets is already being felt. For example, when the 2014 oil price collapse hit, a country like Nigeria experienced its first recession in twenty-five years, with a 1.5 per cent contraction of the economy and a loss of approximately $18 billion (Tänzler et al. 2020). With price volatility continuing, exports declining and the risk of stranded assets growing, the economic and therefore social, political and security sustainability of countries that are dependent on oil and gas revenues and have limited fiscal capacities will be increasingly questioned. With the EU being the first continent aiming to become climate neutral, and with trade relations between the Union and many of these fossil fuel countries being dominated by energy, the onus is on the EU to support their climate mitigation strategies. Were the EU not to do so, its own energy transition would risk exacerbating vulnerabilities, and the political, economic, social and security consequences these would trigger could boomerang back into the Union itself.

Countries like Algeria, Egypt, Iraq, Libya and Nigeria are not only heavily dependent on their fossil fuel exports, but a large share of these go to the EU (see figure 1). They would be severely affected by Europe's energy transition as a result. Libya, whose economy has been shattered by civil war, revolves around hydrocarbons. Around 65 per cent of these go to the EU. Algeria, as the third largest gas exporter to the EU, is almost entirely dependent on it for its hydrocarbon revenues too. These in turn account for 90 per cent of its exports and 60 per cent of its national revenues (Leonard et al. 2021). The EU's energy transition will immediately reverberate on these fossil fuel producing countries' public finances, economies and societies.

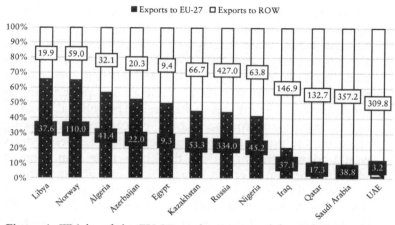

Figure 1: Weight of the EU-27 market in annual fossil fuel exports (2019)

Source: UN-Comtrade.

Note: Units are million metric tonnes. Commodities covered are coal, gas, oil, and petrochemical derivatives.

Moreover, these countries are characterized by deep economic, infrastructural, social, political and security fragilities. Anticipated reduction of exports to the EU would most likely translate into a drop of investment in infrastructure, including for maintenance purposes (Leonard et al. 2021). This is likely to happen already during this decade, even if the EU continues to import approximately the same quantities of oil and gas in the coming years. Beyond the energy sector, Europe's energy transition will affect profoundly the state capacities and state–society relations of these countries. Countries like Algeria and Egypt have very limited financial buffers to shield themselves from price volatility and reduced exports compared, for instance, to the Gulf states (Raimondi and Tagliapietra 2021).[12] They also have bloated public sectors, whereby the state has provided sustenance to large segments of the population, playing a key role in maintaining social and political stability. The lack of democratic representation, political participation and human rights protection has largely been possible over the decades because of the provision of such social services and subsidies 'pacifying' societies. Added to this, the political economy of these countries is intertwined with corruption, clientelism and centralization of political power. As these dynamics are shaken up by Europe's transition, whilst alternative energy and thus governance systems are not in place, political instability, violence and perhaps even greater repression, rather than societal resilience and political transformation could end up being the result. The political, security and economic consequences could be catastrophic. Taking place on the southern Mediterranean shores, these effects would bounce straight back to the EU itself, be it in terms of terrorism, organized crime, smuggling or irregular migration.

Hence, the Union has a key role to play in petrostates that are both vulnerable and dependent on the EU for their energy exports and ensuing economic, social and political stability. While gas will continue to flow to Europe over the next decade, European public and private players can sustain the decarbonization of these countries, by supporting the development of renewables as well as the decarbonization of gas and development of blue hydrogen (Franza 2021).[13] In so far as blue hydrogen plays a temporary role in the EU's own hydrogen strategy, this can provide a key incentive to decarbonize the fossil fuel industries of these neighbouring countries. Algeria has already engaged in carbon sequestration, through its project in Salah, while Egypt could move into this area given its considerable gas reserves, which could be exploited even beyond current production plans. It could also develop renewables and cost-competitive green hydrogen[14] given its solar and wind potential. In Libya, the transition is a harder sell, hampered by deep political divisions, rampant insecurity as well as problems of access to electricity in light of the damaged generation, transmission and distribution assets caused by military conflict and delayed maintenance. However, particularly in view of the latter, the development of solar energy to improve electricity access in remote areas is an attractive proposition, alongside the development of natural gas coupled with CCS.

The aim behind supporting these investments is not that of preventing the exacerbation of vulnerabilities by sustaining the status quo. These energy and governance systems are unsustainable as they are. The aim is rather that of supporting climate mitigation in ways that can transform these countries' economies, energy systems and governance, making them more resilient and effective. This will not automatically mean the transition to more legitimate, accountable, let alone democratic, political systems. Supporting the energy transition does not go that far. However, by supporting the transition of energy systems that are more decentralized, diffuse and dependent on effective hard and soft infrastructures, the EU's climate mitigation policies in fossil fuel producing states can positively influence governance in these countries in ways that can enhance their broader sustainability.

Disarming fossil dependences

A second dimension the EU should account for as it proceeds with its transition is the potential this has to reduce weaponized (and weaponizable) fossil dependences. The past is rife with examples of weaponized energy dependences and the failed potential of energy interdependence to foster peace. Conflicts in the Middle East, the eastern Mediterranean, eastern Europe and the Caucasus amply demonstrate this hard truth. Likewise, relations between Europe and Russia or the Gulf, as well as the influence that these countries have wielded on energy-poor countries in eastern Europe, Turkey, Africa or the Mashraq highlight how energy has been used as leverage to extract strategic, political or economic gains.

Azerbaijan has used its energy wealth built up in the first two decades of the twenty-first century to build its defence sector. The aim was to reverse the outcome of the first Nagorno Karabakh war, which it had lost in 1994. Indeed, a second war was launched – and won – by Azerbaijan against Armenia in the autumn of 2020.

Algeria has also leveraged energy in its conflictual ties to Morocco. Energy used to be one of the few areas that seemed to be insulated from the tensions bedevilling this historically fraught relationship. However, in 2021 Algiers suspended its gas transit through Morocco in what appeared to be an effort to rebalance a geopolitical equation that had tilted in the latter's favour in the Western Sahara, particularly during the Trump administration. These are only a few recent examples of the weaponization of oil and gas dependences. Many more could be cited over the last decades.

Russia, however, represents the most evident and dangerous example of what weaponized fossil dependences imply. Echoing the 2006 and 2009 crises with Ukraine, in 2021 Russia leveraged soaring gas prices and reduced supplies to coerce Moldova's pro-European government to change strategic course. The Russian company Gazprom went as far as pressurizing Chisinau to modify its deep and comprehensive free trade agreement with the EU and delay its energy market reforms, notably those foreseeing the liberalization of gas markets, in exchange for reduced gas prices. Moldova resisted only because the EU came to the rescue with a €60 million grant to withstand the energy crisis.

Russia did not stop there. In 2022, energy became a central piece of its war on Ukraine. As Russia planned its invasion bolstered by rising energy prices, it both contributed to that price spike by reducing its storage levels in Europe and banked on the fact that high prices and Europe's dependence on Russian gas would weaken if not break altogether European resolve to sanction Russia in the event of war. That did not happen, and notwithstanding record high prices, the EU proceeded with unprecedented economic and financial sanctions against Russia. Energy dependence remained a central feature of the war itself. On the one hand, against the backdrop of dizzyingly high prices and the acknowledgement that energy trade with Russia contributed around €1 billion per day to Moscow's war effort, Europeans woke up to the hard truth that eliminating their dependence on Russian gas, oil and coal was a strategic, even existential, priority. On the other hand, Russia itself cut off supplies to Europe and pressurized Europeans to pay for their energy in roubles, in violation of the existing contracts.

Weaning off Russian gas could not be carried out painlessly overnight. On the eve of war, 60 per cent of European imports from Russia were energy products, and European dependence on Russian energy, notably gas, hovered around 40 per cent, with peaks in some Member States reaching almost 100 per cent (Elagina 2021). Severing the EU's fossil ties with Russia implies alternative fossil sources and relationships, building infrastructure – notably for liquefied natural gas – developing a more integrated energy union, as well as accelerating the energy transition. It requires a gigantic effort featuring diplomacy, funds, infrastructure, laws and regulations.

Most fundamentally, it entails shedding two political convictions that permeated the Union for decades. The first and most important is that the energy relationship with Russia could somehow be shielded from geopolitical tension and conflict. Given the energy interdependence – rather than mere dependence – between the EU and Russia, it was believed that both sides would succeed in carving out energy from political and geopolitical turmoil. The history of the Cold War, in which energy trade continued across the Iron Curtain, vindicated this conviction for many decades. The second is the neoliberal belief that within the framework of (light) regulation, the free functioning European energy markets would ensure affordable and secure energy for Europeans.

These two convictions explain why (cheap) Russian fossil fuels dominated the EU–Russian economic relationship up until 2022. The increasingly tense political relationship with Russia over the years did not shake these convictions and ensuing actions. The 2008 war in Georgia, the 2014 annexation of Crimea and war in eastern Ukraine, Russian disinformation campaigns, its election meddling across European countries, and its policies in Syria, Libya, Mali, the Caucasus, Belarus and Kazakhstan were insufficient to alter European energy beliefs and ensuing action vis-à-vis Russia. Likewise, Russia's differentiated gas pricing policies, its recurrent threats to interrupt supplies and its promises of investment in pipeline infrastructure to divide Europeans were also not enough to push Europeans to alter their energy course. True, in the first two decades of the twenty-first century, some initiatives were pursued to reduce EU dependence on Russian fossil fuels, from internal regulations and infrastructure development to the external promotion of diversified energy sources and routes. But these were half-baked successes at best, that did not prevent European dependence on Russian fossil fuels from actually increasing in the decade leading to 2022.

As discussed in chapter 1, the energy security debate, which was prominent in the early 2000s, dropped off the European radar between 2014 and 2021–2. Low oil and gas prices and LNG supplies heading to Europe emboldened the EU and fed the illusion that markets would magically resolve the energy conundrum, making energy insecurity a thing of the past. At the same time, politics kicked in and, especially after the European Green Deal was unveiled, the promise was to ensure European energy security not by buying gas from non-Russian sources and ensuring its smoother circulation within the single market, but rather by reducing the overall amount of gas consumed within the EU.

The global and especially European energy crisis triggered by Russia's invasion of Ukraine profoundly questioned European convictions and brought about a policy shift that – uneasily yet necessarily – blended pragmatism and principle. Pragmatically, the war in Ukraine emphasized the fact that Europeans had to reduce their fossil ties with Russia immediately, and the only way to do so was by developing other fossil relationships with producer countries in Africa, the Gulf, the Caucasus, as well as the US. At the same time, standing firm on principle meant that stopping there would be strategically and normatively myopic. Diversification of fossil relationships had to come alongside an acceleration of the energy transition. European energy independence achieved through the short-term diversification of fossil relationships with third countries, their medium-term greening, and the EU's long-term decarbonization became not just a normative vision for the future but a strategic imperative here and now.

In Europe, a more decentralized and technological decarbonized energy system supports the EU's quest to become strategically autonomous (Tocci 2021a). Whereas new dependences and interdependences are arising – notably concerning critical minerals – these are not comparable to those existing in a fossil-fuelled Europe. However, as discussed at length in chapter 4, pursuing European strategic autonomy in the energy transition should not be interpreted as a drive to sever interdependences *tout court*, but rather as an effort to eliminate weaponized (or weaponizable) dependences both by diversifying fossil relationships and greening them over time. The war in Ukraine made Europeans realize this vis-à-vis Russia. However, the argument applies across the board, with regard to relationships with producer states in Africa, the Middle East and the Gulf.

At the same time, the severing of the EU's fossil relationship with Russia is a strategic necessity that should not lead to turning a blind eye towards its strategic downsides, and planning ways to mitigate these when political conditions arise. The clear value of reducing the EU's dependence on Russia's fossil fuels risks being dampened by the geostrategic cost of pushing Moscow further into Beijing's lap. Even before the war, this was the trajectory of Russia–China relations. Until recently, Russian gas exports to China were marginal, and the EU dwarfed China as a destination market for Russian gas. However, the Power of Siberia gas pipeline opened in December 2019 and started changing the equation, with its expected 38 bcm of gas per year to China eventually representing around 15 per cent of Russian natural gas export volumes.[15] Even more important given it would imply a diversion of resources from Europe to Asia, progress in the SoyuzVostok 'Power of Siberia 2' gas pipeline via Mongolia, with its expected 50 bcm of capacity, means that close to 100 bcm of Russian gas – excluding LNG from the Arctic – could head to China (Gazprom 2021). Even before the war, Gazprom's industrial plan explicitly mentioned the objective to increase its exports to Asia from 5 per cent to 30 per cent. Coming alongside closer economic, political and military ties between China and Russia, the severing of EU–Russia fossil interdependence, while strengthening European energy security, comes with heavy geostrategic baggage.

If and when political conditions allow for a resumption of EU–Russia cooperation, the Union's aim should be to pick up the pieces of its broken fossil interdependence with Russia and chart a decarbonization trajectory between the two, not least given Russia's potential for the development of renewables and decarbonized hydrogen. Much like the US acknowledges that its competition with China stretches across all policy areas apart from climate where cooperation should prevail, the same logic should eventually apply to the EU and Russia when political conditions enabling engagement are restored.

Addressing the direct effects of EU transition policies: the Carbon Border Adjustment Mechanism

A third space to watch concerns the direct effects of EU climate policies. The most poignant case is that of the Carbon Border Adjustment Mechanism (CBAM). Against the backdrop of increasing carbon prices in the EU, the 2021 'Fit for 55' package put forward the proposal for a CBAM: a carbon levy imposed at the EU's borders to reflect the embedded carbon content of certain goods imported from third countries that do not penalize emissions. Whereas imposing a CBAM on all goods would be technically difficult – if not impossible – and politically unsustainable given the barrage of resistance the Union would face, Brussels wisely picked selected industries. A CBAM would thus be applied to the iron, steel, cement, fertilizer, aluminium and electricity sectors, i.e., areas in which greenhouse gas emissions are typically high and relatively easier to measure.

Inbuilt in CBAM are three objectives. First, the carbon border tax aims to prevent European companies from relocating their emission-intensive activities beyond the EU's borders, a phenomenon known as 'carbon leakage'. With emissions and climate change knowing no borders, this would do nothing to advance the global public good of decarbonization. Second, CBAM aims at avoiding a competitive disadvantage for European industries that are subject internally to rising carbon prices, the cancellation of free emission allowances and eventually an extended remit of the ETS, compared to external competitors that are not. Whereas many countries are moving towards different forms of carbon pricing, very few feature the same scope, complexity and ambition as the European ETS. Given that an expanded scope and lowered caps of the EU's ETS are prime features of the European Green Deal, the European Commission must avoid these unfairly harming European industry. Particularly at a time when industrial policy has become fashionable after decades of neglect, and the Union strives to decouple emissions and economic development, a climate policy that ends up destroying European industry would be paradoxical to say the least (European Commission 2021g). However, this objective should not be read as an ambition to preserve European industry as it is. Simply allowing the continuation of free emission allowances, and through these preserving industrial revenues and jobs, would have done the trick. The point is rather that of decarbonizing both production and consumption by exposing emission-intensive European industries to a higher cost of emitting in a way that induces their transition rather than eliminating them altogether. Paired with a revised ETS, and ideally with measures aimed at not harming European exporters,[16] CBAM is about fostering Europe's transitioning industries by ensuring a level playing field with third country peers that would otherwise not be subject to the same costs of emitting.

These first two objectives make CBAM necessary. They are also what make CBAM probably compatible with the rules of the WTO. These require countries to avoid discriminating between domestic and foreign goods, and between foreign goods themselves, as well as ensuring that trade restrictions are proportional to the environmental goals pursued. Ensuring WTO compliance was indeed a main driver in the design of CBAM,[17] explaining, inter alia, why the Commission's proposal did not feature exemptions for developing countries. Exemptions of any kind, including for developing countries that would be adversely affected by CBAM, may well make CBAM vulnerable to successful challenge at the WTO. Compensatory measures for such countries should be devised, notably to support their transition journeys, but beyond the scope of CBAM itself.

Alongside these two necessary objectives is a third objective, which makes CBAM not just necessary but also desirable from a foreign policy perspective. Embedded in the wider notion of the EU as a climate leader, able to exert both the ideational power of example and material leverage given its economic weight in the world, CBAM is aimed at triggering virtuous climate policies beyond the EU (Vanheukelen 2021). Whether these take the form of cap-and-trade market-based mechanisms, carbon taxes, regulations or funding instruments does not matter, so long as such policies are effective in inducing decarbonization. The very idea of CBAM – well before its eventual implementation in 2026 – is expected to contribute to the energy transition beyond EU borders, particularly in countries that would otherwise be adversely affected by it. CBAM is the feature of the European Green Deal that has attracted most international attention and to some extent, as we shall see below, third country responses. This suggests that in some ways this third objective is pursued already, even before the implementation of the levy itself.

The European Commission's CBAM proposal has been criticized in Beijing and has raised eyebrows in Washington. I will return to this, and how dynamics underpinning carbon pricing interlock with the broader geostrategic rivalry between the US and China, in chapter 4. CBAM has also been critiqued from a development perspective, given the concern that it would adversely affect the development prospects of the Global South (Weko et al. 2020). In particular, the worry is that developing countries with high shares of carbon-intensive exports to the EU would be exposed to additional costs, and therefore deteriorated terms of trade and reduced export shares. Given the financial and technological limits hindering the transformation of developing countries' industries, as well as their inability to measure and account for the carbon content of their exports, it should be the EU's responsibility to support the development of such capacities, in turn facilitating either their eventual exemption from CBAM or its reduced impact.

However, delving deeper reveals that the countries that could be most affected by CBAM, given both the potential impact of the levy and the relative weight of the EU in their trade relations, are middle-income nations located in the EU's surrounding regions (see figure 2). The EU has significant and often highly problematic relations with several of these countries. This increases the risk, amongst other things, of retaliation. In other words, if amidst tense economic and political relations, CBAM is depicted by affected third states as trade protectionism in disguise, it might trigger tit-for-tat retaliation. This is particularly true in a world in which protectionism is already on the rise and is politically popular amongst the public both within and beyond the EU.

Setting aside China and Ukraine, which will be discussed below, the UK – whose own ETS would likely exempt it from CBAM – and Russia – amply discussed above – Turkey is the country that would be most seriously affected by CBAM if the levy were implemented today. This is due to the high

Exports to the European Union 2019 in selected sectors to be considered in the CBAM. 20 most-exposed countries in terms of aggregated value of exports (billion $)

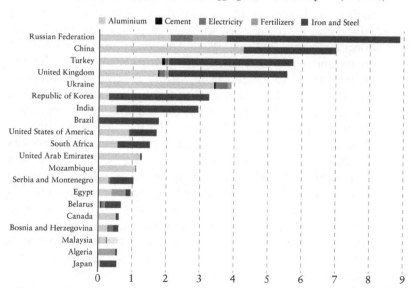

Figure 2: Potential impact of CBAM on the EU's trading partners

Source: from UNTCAD News ('EU Should Consider Trade Impacts of New Climate Change Mechanism', 14 July 2021), by UNTCAD, © United Nations. Reprinted with the permission of the United Nations.

intensity of emissions of its main exports to the EU. From Turkey, the EU imports large amounts of steel, cement and to a lesser extent aluminium and paper. CBAM could add fuel to an already highly complex relationship.

EU–Turkey relations are highly strained (Tocci and Aydın-Düzgit 2015). For well over a decade now, academics, practitioners and pundits have deplored Turkey's drift away from the West, its domestic slide towards authoritarianism, and its growing assertiveness and independence in foreign policy, including Ankara's warmth towards Vladimir Putin's Russia and visible distancing from the EU and the US (Tocci 2021b). Long gone are the days of unambiguous partnership, integration and friendship: at most Turkey and the EU can be defined as frenemies, ready to cooperate when the occasional interest overlaps but invariably looking at one another with palpable mistrust, perhaps even dislike (Cook 2017). In fairness, Turkey is not the only one to blame. At least since 2005 – i.e., since Turkey began EU accession negotiations – the Union has been all consumed by its successive internal crises, and its bandwidth for foreign policy has diminished. Specifically, the willingness to engage in further rounds of enlargement has shrivelled, being further damaged by the evident de-democratization in former enlargement countries like Poland and Hungary. Given the difficulty of ensuring that democratic standards are respected after a country enters the EU, the general willingness to let new members into the club has plummeted even amongst the staunchest supporters of enlargement. Enlargement, beginning with Turkey, has been shelved for the time being. Alas, it has not stopped here. As Turkey de-democratized and began antagonizing EU Member States Greece and Cyprus, the European debate on Turkey, far from focusing on integration, has revolved around sanctions instead. Turkey's relations with the EU dangerously teeter at the brink, with the EU periodically returning to the question of whether and how to sanction Turkey.

The European Green Deal and the prospect of CBAM are woven into this bleak picture. Turkey has followed the development of CBAM with concern. CBAM indeed presents two sets of problems for Turkey. First, as for most neighbours, the EU represents Turkey's main trading partner, with over 40 per cent of Turkish exports destined for the EU (European Commission 2021h). Much of this is carbon-intensive trade, suggesting that CBAM could end up costing Turkey up to $900 million per year (Fleeson 2021). Second, Turkey is part of the EU customs union, whose functioning could be severely hindered by CBAM (Ulgen 2021). A customs union entails the free circulation of goods, which is ensured, inter alia, by eliminating controls of origin certificates. The Commission's CBAM proposal requires a calculation of the carbon intensity of imports to determine the corresponding carbon levy. This requires knowledge of the origin of goods. Imposing such a requirement on Turkish imports would depart from the current customs union arrangements, hampering its functioning. In practice, it could end up transforming the EU–Turkey customs union into a mere free trade arrangement, in which origin documentation is necessary. By contrast, if origin certificates continued not to be required after CBAM comes into effect, third countries with higher carbon tax rates than Turkey could deflect their trade and re-export their goods via Turkey to the EU, undermining the goal of preventing carbon leakage (Ulgen 2021).

The challenge here is therefore to use the years ahead – before CBAM enters into force – to devise ways of including Turkey into the remit of the European Green Deal, exempting it from CBAM. This could happen either by including Turkey in the European ETS like Norway or by linking a Turkish ETS to the European one, like Switzerland. In both cases, Turkey will need to establish a meaningful domestic carbon pricing system, which will not be easy. The EU and Turkey are at very different stages of their transition journeys, with Turkey having done little until now. However, in 2021 Ankara did embark on its energy transition, finally ratifying the Paris Climate Agreement and announcing its goal of climate neutrality by the 130th anniversary of the Turkish Republic in 2053. The European Green Deal, including the threat of CBAM, represents a powerful incentive for Turkey to take its first green steps to maintain the competitiveness of its export-driven economy. Today, these are little more than aspirations, with no plan to get from A to B. But this is where Turkey's relationship with the EU could come into play.

The EU has held out the prospect of modernizing the 1996 EU–Turkey customs union, extending it to areas such as services, procurement, state aid and trade dispute settlement. Talk about modernizing the customs union has been ongoing for several years now, making a 'modernization' of the 'customs union modernization agenda' necessary. In particular, given the EU's own evolution in the meantime, an upgraded customs union with Turkey today cannot exclude the green and digital dimensions. Ankara's awakened decarbonization aspirations, the prospect of a modernized customs union and the spectre of CBAM could thus be constructively combined to kick-start Turkey's energy transition. In particular, 'greening' the modernized customs union agenda could enable green technology transfers, support Turkey in the development of renewables and the governance that goes with them, improve the country's energy efficiency as well as further liberalize its power market (Aydıntaşbaş and Dennison 2021). All this, plus the fact that up to €1 billion of EU pre-accession funds to Turkey could be earmarked for climate priorities over the seven-year budgetary cycle, could jumpstart Turkey's transition.

In cases like Turkey, and more broadly in emerging and developing economies that could be negatively affected by CBAM, the success of the EU's carbon levy hinges on what happens before it eventually comes into effect. Put bluntly, CBAM will be most successful if the need for it diminishes to the point of disappearing altogether by the time it is expected to be operationalized. If operationalized, CBAM, while addressing carbon leakage, could severely harm countries that have significant yet carbon-intensive trade with the EU. It would also harm developing countries with high carbon lock-ins, unclear decarbonization paths and limited statistical and administrative capacities to measure, verify and report the carbon content of their traded goods (Eicke et al. 2021). In a world in which protectionism, conflict and violence are on the rise, CBAM could also spark future trade wars. This said, CBAM is an essential piece of the European Green

Deal puzzle. Its real success hinges on whether the prospect of CBAM will serve as an additional incentive for neighbouring countries to accelerate their own decarbonization. If this happens and if the EU steps up its support to facilitate these transition journeys, including by using international carbon dividends from CBAM to increase the climate funds channelled to the least developed economies, CBAM's success will have become the cause of its own demise. A CBAM with no proceeds is ultimately what the aim of CBAM is, or should be, about.

Unlocking transition opportunities in neighbouring regions

While addressing the vulnerabilities arising from the climate crisis, European foreign policy must also unlock the green opportunities that could emerge from the energy transition to strengthen state and societal resilience beyond its borders. This is a fourth dimension to explore when addressing the geopolitical implications of the EU's energy transition in neighbouring regions.

As laid out in the IRENA (2019a) report mentioned above, a decarbonized world could be more prosperous and peaceful given the more abundant and decentralized distribution of renewable energy sources compared to the geographically concentrated and more limited presence of fossil fuels. The greater availability and more even distribution of wind and sun compared to fossil fuels imply potentially reduced competition over natural resources, fewer weaponizable energy dependences, and a more diffuse distribution of energy power in the international system. This 'democratization' of energy, so the argument goes, will foster prosperity and peace. Applying this logic to Europe's surrounding regions, which is precisely where fossil sources are most heavily concentrated, makes for a compelling case. Two categories of countries stand to benefit the most in this respect, and it is up to the EU to support them.

The first category is the oil and gas transit states, which, while energy poor themselves, have enjoyed significant transit fees to which they became addicted. The case of Ukraine stands out. Ukraine's energy security has long been a dominant feature of its domestic debate and a priority in relations with the EU. Part of the Energy Community[18] since 2011, Ukraine has strived to integrate its energy market with that of the EU for years (Pirani 2021). The ambition to strengthen EU–Ukraine energy cooperation increased with the 2016 EU–Ukraine strategic energy partnership, with a view to embedding Ukraine in all dimensions of the EU's energy union at the time. The energy partnership featured grants, technical assistance and investments in energy efficiency as well as work on the flexibility of the Ukrainian power system and its synchronization with the European electricity grid. That work is yet to be completed, and Ukraine's regulatory harmonization and infrastructural connectivity with the EU remain key to enhancing the country's energy security (Bielkova 2021). It was during the war in 2022 that the EU moved decisively to connect Ukraine to the European electricity grid.

However, both the EU and Ukraine are well aware that Ukrainian strategic resilience requires much more: weaning off Russian gas transit fees, which gave Moscow leverage over Kyiv for many years. Ukraine has been a key transit state for Russian gas to the EU and feared losing the revenues derived from this role. In retrospect, this may look like evidence of a Stockholm syndrome but it was simply the reflection of the country's deep economic fragilities. With 40 per cent of Russian gas exports transiting through Ukraine under the 2019 agreement reached with Moscow and brokered by the EU, Kyiv received around $3 billion per year in fees (Bond 2021). Losing these revenues was difficult to swallow so long as Ukraine remained dependent on Russian gas, making the country even more vulnerable to Moscow. In other words, Ukraine's long-term strategic resilience required it to wean itself off Russian transit fees, but this contradicted its short-term economic and energy resilience given the deep fragilities of the country.

The war has made the reconciliation of this contradiction imperative. Eventually, the only way to square the circle is by developing Ukraine's green industrial capacities. Only this can eliminate structurally Moscow's potential to arm-twist Kyiv. Already prior to the war, Ukraine aimed to increase its share of renewables to 25 per cent of its energy mix by 2035 (Prokip 2019). The European Green Deal offered the potential to accelerate significantly this journey. In 2021, the Commission launched an initiative featuring twinning, reskilling and funding to support the transition of Ukraine and the Western Balkans away from coal (European Commission 2021f). Furthermore, Germany, together with the US, committed to establishing a green fund for Ukraine to jumpstart investments in renewables, hydrogen and energy efficiency, aiming to mobilize $1 billion worth of investments (US-DOS 2021a). Ukraine's offshore wind potential is particularly promising in this regard (Sabadus 2021). It is crucial that these investments materialize as part of Ukraine's reconstruction when this gets underway, with an aim not simply of exporting Ukraine's clean energy to the EU – as suggested by the European Commission's hydrogen strategy – but above all to decarbonize Ukraine's own highly insecure and carbon-intensive energy system. This is a strategic priority for the EU and Ukraine alike.

Another category of countries that stand to benefit from the energy transition are fossil fuel-poor countries with a high potential for the development of renewable sources. The green transition in North Africa for instance can both help domestic development as well as create export opportunities to Europe that did not exist in a fossil fuel-only energy system (Franza 2021). Sceptics would retort that, while great on paper, precedents suggest that this is unlikely to happen in practice (Schmitt 2018). Indeed, ambitious regional projects in Europe's neighbourhood have never seen the light of day. The mammoth renewable energy project Desertec is probably the most infamous example. Conceived in 2003 by the Club of Rome as the Trans-Mediterranean Renewable Energy Cooperation and launched in 2009 as the Desertec project, the goal was to export electricity to Europe generated by solar and wind sources on the edges of the Sahara Desert. The project never came to fruition, tripping over infrastructural, institutional, economic and geopolitical hurdles. Many of these problems persist, ranging from regulatory barriers and inadequate support schemes to difficult access to capital and insufficient electric infrastructure. Large regional initiatives are unlikely to bear fruit to this day. While some renewable projects have been implemented on Europe's southern shores, notably in Morocco, these are limited in scale with only residual export capacity to the EU, not least given the very limited interconnections between the EU and its southern neighbours.

This said, country-specific energy transitions in Europe's surrounding regions, coupled with a lighter regional touch, are more likely to deliver today than in the past because of a profoundly changing energy context in Europe itself. It is the EU's own transition that creates a fundamentally different set of hard and soft incentives for change beyond its borders. More concretely, if the EU is to meet the renewables targets it set for itself in the 'Fit for 55' package, and also jumpstart its hydrogen economy (European Commission 2020b), working with its neighbours is essential. Put bluntly, Europe's population density coupled with NIMBY resistance means that Europe neither has the physical nor the political space to install all the renewable capacity the EU foresees. Much of this would need to be installed in the southern Mediterranean, where the potential for renewable sources is particularly high. Today, the renewable potential of Europe's surrounding regions remains largely untapped, while the development of hydrogen is an interesting possibility, which the EU's own transition could help unlock. This is not to say that all such clean energy would be exported to Europe. It would also be essential for local economic development. One only needs to think of the role that domestically produced green hydrogen could play in Morocco's fertilizer industry that currently relies heavily on imported ammonia and thus on (grey) hydrogen as feedstock (Franza 2021). Renewable energy and green hydrogen produced in the southern Mediterranean could both support local economic development, as well as be partly exported to Europe in ways that might be technically feasible and cost competitive, not least due to the existence of repurposable gas pipelines and the potentially low costs of electrolytic production in North Africa (Franza 2021).

Complex crises and EU foreign policy

A final dimension to be considered turns the question on its head: rather than the impact of the EU's energy transition on its neighbours, it focuses on the impact of the climate crisis – and thus of insufficient transition – to date. The link between the climate crisis and European foreign policy is a topic whose relevance has steadily grown since it was first raised in 2008 by the High Representative (European Council 2008). The exacerbating climate crisis is felt by states and societies in the EU's surrounding regions in ways that affect profoundly both the EU's climate policies as well as other foreign policy areas, from development to migration, security and defence (Werz and Hoffman 2016). This is taking place in regions that already suffer from profound fragilities, revolving around development, governance, security and human mobility. The Sahel and Iraq are two cases in point.

The Sahel has been afflicted by profound instability for a long time. Internal conflict, transnational terrorism, and the trafficking of arms, drugs and people have destabilized the region. Nigeria has long endured insurgency in the Niger Delta, alongside religious and ethno-political conflict in the north. Niger has witnessed conflicts over water and land use, and competition over scarce natural resources has fuelled tensions. The 2007–9 Tuareg rebellion is a clear example. Algeria, Mali and Nigeria are plagued by violent extremism and terrorism, with areas of limited statehood providing fertile ground for such groups to thrive (Boas and Strazzari 2020). Since 2001, there have been over 1,000 terrorist attacks in Algeria alone. The pandemic, including its health and socio-economic impact, has added fuel to the fire. Tunisia's democracy, which was championed as the last hope of the now defunct Arab spring, has now flipped back in democratic recession, with the government's catastrophic handling of Covid-19 becoming the straw that broke the parliamentary democracy's back, ushering the way back in for authoritarian governance.

All this connects with human mobility patterns, which have seen North Africa and the Sahel emerge as a crucial hub for migration flows headed north towards Libya and across the Mediterranean to Europe. The Sahel has evolved into being an origin, transit and destination region at the same time. This has been true of Morocco for many years and is becoming true also of countries like Niger and Mali. Inadvertently, these countries have become both origin and transit states for African migrants headed north as well as transit and destination countries for migrants expelled from Libya and Algeria. Whereas in 2003 an estimated 65,000 sub-Saharan migrants travelled north across the desert via Agadez in Niger (Di Bartolomeo et al. 2011), in 2019 the International Organization for Migration (2020) registered over 370,000 outgoing and incoming persons. This includes Nigerian migrants moving north towards Algeria and Libya, as well as those returning to Niger.[19] One year later, despite Covid-19, the flows persisted unabated. All this has led, inter alia, to a significant rise in food prices and ensuing food insecurity.

So far, the EU has responded to the complex crisis of the Sahel through a joined-up approach that blends security and development policies, alongside an increasingly pronounced focus on migration. The security–development nexus was first put into practice precisely in the Sahel, beginning in the early 2000s (Faria 2004). It gradually evolved into an integrated approach in the 2010s, in which civilian, military, development and migration policies and players started working together (Ahmed et al. 2018). The combination of security sector capacity building, development policies and cooperation with African regional and sub-regional groupings like the African Union and the G5 Sahel transformed the Sahel into a laboratory of experimentation of European foreign policy (Venturi 2017).

While seeking to be joined up, the EU's Sahel policy is far from effective, however. In the wake of an ever more securitized migration debate in Europe and an ensuing political stalemate within the EU over a common migration and asylum policy, the Union began outsourcing its migration governance, notably to countries like Niger. External migration management became the surrogate for the EU's failure to agree on an internal migration and asylum system. This has delivered localized results: EU–Niger migration cooperation, for instance, has reduced migration flows through Niger itself. On a much larger scale, the same is true of flows through Turkey, in light of the 2016 EU–Turkey migration deal. However, in the absence of an overall system featuring EU external, border as well as internal migration and asylum policies, the net migration balance between Europe and Africa has remained largely unaltered. EU–Niger migration cooperation seems to have diverted flows elsewhere, with Malian routes becoming increasingly popular.

This has generated two sets of problems. On the one hand, the reduction of flows through Niger has affected all those local livelihoods that revolved around the migration economy, including the selling of food and lodging for migrants. With no alternatives to replace these, the result has been the impoverishment of already vulnerable individuals in the country.

On the other hand, the diversion of flows through Mali has exacerbated the growing fragilities in that country. Indeed, in the meanwhile instability in Mali has been rising. Against the backdrop of two successive coups in 2020 and 2021, French President Macron announced the winding down of operation Barkhane, the military endeavour launched upon invitation of the Malian authorities in 2013 in view of the 2012 coup and the growing threat of terrorism in the country. With Malian security forces unable or unwilling to take on greater responsibility, French public opposition against a 'forever war' in the Sahel grew, echoing the US drive to leave Iraq and Afghanistan. Rather than pulling back from the Sahel altogether, Paris sought to Europeanize its military involvement there, achieving mixed results. Faced with the political, institutional and legal hurdles bedevilling a more muscular EU defence involvement – namely an EU military operation – France led a coalition of the willing between Member States – the Takuba Taskforce – to deal with the growing threat of terrorism in the region (Lebovich 2021). This happened amidst an increasingly complex global dimension of the Sahel conundrum, with the Russian Wagner mercenary force filling the security vacuum in Mali. Eventually both France and the Takuba Taskforce wound down their military operations in Mali. The Sahel continues to be an area in which the EU and its Member States are deeply involved. However, Europe's joined-up development–security–migration policy is full of holes, not delivering tangible long-term results to date.

Against a backdrop of deep fragility, violent conflict and multifaceted European involvement that is yet to deliver long-lasting outcomes, climate change encroaches, making crisis even more complex. In the years ahead, around three quarters of the Sahel's arable land currently used for agriculture will suffer from droughts (Venturi et al. 2020). Nigeria, for one, is losing 3,500 square kilometres to desertification per year, while hundreds of villages have already disappeared in Niger. Niger, northern Nigeria and south Algeria all experience ever more frequent droughts and floods, while Algeria and Morocco's northern coasts are threatened by rising sea levels and land erosion. The deepening climate crisis in the Sahel will increasingly interlock with and exacerbate existing fragilities, revolving around weak governance, socio-economic vulnerability, insecurity and human mobility (Werz and Conley 2012).

The EU, much like other international players, acknowledges the Sahel's climate crisis. Its Sahel strategy accounts for the region's environmental degradation as well as for the risk of such degradation exacerbating instability and conflict (EEAS 2011, 2021). This means that the EU's climate adaptation funds will need to grow significantly.

However, simply increasing the funds devoted to climate adaptation is not going to work. Just as important is rethinking how such climate funds are deployed. The link between climate change, security and migration is far more complicated than first meets the eye, and EU policies have failed to account for this fully, at times inadvertently aggravating drivers of instability and conflict (Raineri 2020). Weak, ineffective and often unaccountable governance plays a major role in explaining both migration and insecurity as well as the damage caused by exogenous factors such as climate change.

As the climate crisis escalates in the Sahel, climate adaptation must become a prime focus of EU foreign policy. Today, while playing an important role in EU conflict prevention activities, climate still represents an appendix of foreign policy, including security and defence, often being described as a multiplier of existing crises and thus of EU policies towards them. Rather than climate, the management of migration, or rather the containment of migration from and through the region, is in the political spotlight. With migration being the *primus inter pares* priority of EU foreign policy in the Sahel, the EU's development programmes are skewed towards funding areas that address the 'root causes' of migration, i.e., factors that inform migration decisions such as job opportunities over and above equally important areas such as infrastructure development or the provision of health and education services. The time horizon of these 'root cause' priority areas tends to be rather short term, with all eyes set on their potential impact on year-on-year flows. These areas do not always overlap with those identified through a sustainable development lens, including a climate adaptation one, which are generally longer term in nature. In fact, a sustainable development lens would see migration, not simply as a phenomenon to be contained, but also as a climate adaptation strategy in and of itself (Raineri and Rossi 2017).

When elaborating sustainable development partnerships with Sahelian countries, the EU must be prepared both to listen more to local priorities and to offer deals that include substantial climate adaptation funds, deployed alongside trade, development assistance, governance, health and security sector support. This also means that climate adaptation funds should not be granted as blank cheques to authorities in the region. Many of the fragilities, including those exacerbated by climate change, are man-made, and therefore simply channelling more money to central governments for climate adaptation will not automatically do the trick. Ramping up the attention paid to climate adaptation, over and above migration management, would include a greater EU focus on work with local authorities, putting a premium on the role played by EU delegations in the region (Goxho and Mourier 2021). This would include scouting for and supporting local climate adaptation initiatives that feature an inclusive and bottom-up approach to natural resource management (Venturi and Barana 2021: 10). A stronger focus on climate adaptation would also include greater engagement on these themes with sub-regional organizations such as ECOWAS (Puig Cepero et al. 2021).

Iraq is another country that has been in the throes of complex crisis for decades. Since the 2003 US-led invasion, Iraq has been torn by sectarianism, limited statehood, terrorism and regional rivalry between Iran and Saudi Arabia. Added to that, it is undergoing a growing demographic crisis, with its current 38 million citizens expected to balloon to 80 million by 2050. Amidst these multiple vulnerabilities, climate change and the energy transition risk bringing Iraq to its knees. On the one hand, like all petrostates, Iraq must embark on its energy transition, whilst lacking the financial reserves to weather the storm. Despite its significant natural resources, Iraq stands out for its mismanagement of them, ranking as the second worst country worldwide for gas flaring (World Bank 2021). This has both created significant environmental and climate damage, and led to the paradoxical situation in which Iraq has been unable to supply sufficient energy to its own citizens, importing gas and electricity from Iran.

On the other hand, Iraq is ranked fifth worldwide in terms of being affected by rising temperatures, territorial degradation and water scarcity (UNDP 2021). Average temperatures are expected to increase between 3 and 5°C compared to 1960 (Mesopotamia Revitalization Initiative

2021). Years of below-average rainfall have made Iraqis ever more dependent on the Tigris and the Euphrates rivers, whose flows to Iraq have been squeezed to a trickle by dams upstream in Turkey and Iran (Loveluck and Salim 2021). Higher temperatures and less rain represent existential threats to Iraq's agriculture, heightening food insecurity and endangering tens of thousands of jobs (Ahmad and Renade 2021). Furthermore, sea level rise as well as water scarcity are driving internal and international migration, with up to seven million people at risk of displacement (Mesopotamia Revitalization Initiative 2021).

The climate crisis in Iraq is already acute, but climate adaptation has hardly begun, including work to climate-proof agriculture and infrastructure as well as to provide urban adaptation options in coastal areas. The first serious attempt to focus on the crisis came in the autumn of 2021 with the Mesopotamia Revitalization Initiative, on the basis of which Iraq elaborated its nationally determined contribution for COP26. The initiative focuses on reforestation, agricultural modernization, water and waste management systems, energy efficiency, circular economy, the capture of fugitive methane emissions and solar energy. While late in coming, it represents a first step to address the escalating crisis.

The EU must do much more to support Iraq in this context. The Union is present through its diplomacy and development assistance, alongside its civilian mission to support the reform of Iraq's security sector. Here too, albeit in a more limited way compared to the Sahel, the EU promotes an integrated approach that blends development, diplomacy and security policies. Just like in the Sahel, however, impact is limited, and climate is an afterthought at best. Whereas the EU is engaging in projects aimed at climate mitigation, notably the modernization of the energy system with a view to reducing natural gas flaring and improving energy efficiency, its work on adaptation is embryonic. If the EU is to play a role in containing the climate crisis in Iraq, with its spill-over effects on development, migration and security, taking on board the climate areas identified by the Iraqi government is a good place to start. At the same time, international energy companies, including European ones that are heavily invested in Iraq, have a large role to play in spurring renewable energy projects as well as eliminating venting, reducing flaring and capturing fugitive emissions in their fossil projects in the country, which account for 40 per cent of Iraq's greenhouse gas emissions.

In neighbouring regions disproportionally affected by the climate crisis, of which the Sahel and Iraq are but two, the EU's focus on climate adaptation must rise steeply up the list of foreign policy priorities. Its climate adaptation strategy (European Commission 2021e) recognizes as much. It calls for an uptick in climate adaptation funds and support for countries in their policies to induce climate-resilient investments and nature-based solutions. It also warns of the need to design adaptation initiatives in conflict-sensitive ways. The strategy, however, is largely focused on internal European measures, reflecting the overall internal bias of the European Green Deal. Elevating its insights into priority areas of EU foreign policy in its surrounding regions will be key.

* * * * * * * * * * *

All the issues and countries discussed in this chapter point to the same conclusion: the development of green capacities in the EU's neighbourhood and beyond. Whether the focus is on climate adaptation, climate mitigation, particularly in fossil fuel producing or transit states, on greening interdependences or on unleashing green economic opportunities, the bottom line is the need for the EU to refocus its external action by putting energy and climate at the core.

The secret behind the Paris Agreement's success was its switch from top-down prescriptions to bottom-up, voluntary and non-punitive commitments. Having set the global ambition to contain warming to 1.5°C, countries were given a free hand in setting their own decarbonization targets – their nationally determined contributions (NDCs) – and their plans to meet them. In many developing countries, these were patched together too quickly (UNDP 2021), with wide gaps in strategy, let alone prospective execution.

As discussed above, apart from the EU, concrete plans to meet net zero emissions have been hard to come by. In some cases, there is insufficient willingness to move in this direction. But in many cases – notably in Europe's surrounding regions – countries mostly do not have the technical, financial and governance capacities to do so. In its quality assessment of NDCs, the UNDP (2021), while signalling high above average scores on robustness, inclusivity and ownership, highlighted that only 27 per cent of assessed NDCs scored above average on feasibility. The difficulty in attracting financial, technological and technical support underpins the difficulty of translating plans into practice. This obviously connects to the broader debate about climate finance and the need for developed countries to support the more vulnerable in this regard. I will return to this in the next chapter. However, it also concerns the need to induce technology transfers and governance reform. In many cases, countries lack the institutional capacities to ensure the necessary data collection, management and communication necessary for them to make meaningful climate pledges, let alone to deliver on them. With the decision taken at COP26 to increase the frequency of reporting of countries' NDCs, whereby these are now due every year, the need for institutional capacity building has become even more acute.

Europeans are well aware that climate action and the energy transition are global public goods, and the EU knows in theory that its European Green Deal will be a success only if it spills beyond its borders. However, in practice the focus of the EU's green agenda is internal. While this is key, it is insufficient. It is high time for a green Europe to start walking the walk of going global.

4

A Green Europe amidst Global Rivalry

Europe's energy transition will reverberate across its surrounding regions, in some cases exacerbating socio-economic, security and governance fragilities, in other cases generating new economic and political opportunities. It will also change regional interdependences, with the geopolitical consequences these may entail. It is essential to address these regional ramifications, both through a more developed external dimension of the European Green Deal, as well as by mainstreaming energy and climate in all dimensions of EU foreign policy.

Given the EU's role in the world, notably in the economic, technological and financial domains, its energy transition will affect not only its own region, but broader global dynamics too. This chapter unpacks the global ramifications of Europe's transition, especially in its natural resource, industrial and financial dimensions. It analyses how the industrial implications of the EU's transition, including the critical minerals this requires, as well as its policies in areas such as carbon pricing and sustainable investments, will influence profoundly relations with China and the US, as well as the global rivalry between them. In doing so, the EU's transition will affect deeply the changing nature of globalization in the twenty-first century. EU energy and climate policies will also have an impact on multilateral cooperation, and shape ties with the Global South, particularly in areas such as climate finance.

A new bipolarity

The world is fast settling into a new bipolar structure, largely revolving around the US and China and the political systems and values they respectively represent (Allison 2017). Under President Biden's administration, in fact, the strategic rivalry between the US and China is no longer framed in purely transactional Trumpian terms, and reduced to a list of sectoral disputes, be it about tariffs, 5G, maritime security, cyber or arms control. These tensions and disagreements are grouped together under a common ideational umbrella: it is a conflict between political systems and ideologies (Brands 2021); democracies versus autocracies. Unlike the post-Cold War international liberal order and the US hegemony it rested on when the key question was the promotion of liberal democratic values, the main question today is the protection of such values in liberal democracies. Liberal democracies strive to protect their values against the direct and indirect attacks of authoritarian powers as well as illiberal nationalist populist forces at home.

In this respect, global confrontation in the twenty-first century harks back to the twentieth-century Cold War and the ideological competition between capitalism and communism. That too was a competition that revolved around contrasting political systems. Unlike the Soviet Union, China does not (yet) have an explicitly international project. Rather than promoting its political system abroad, the Chinese regime appears to be primarily focused on retaining control at home (Mitter 2021). Chinese foreign policy seeks to contribute to this goal by securing friendly third country attitudes towards Beijing. But whether such countries are governed by democracies or autocracies seems less important. What matters is the extent to which they enable China's economic development, while not interfering in its domestic affairs.

However, like the Cold War – and arguably more acutely than it – today's global competition revolves around proving that one's model of governance – liberal democracy versus authoritarian capitalism – is superior to the other, be it in terms of prosperity, technology, security or public health.

Whereas the deep current of this new bipolarity is exquisitely political in nature, the surface waves at which the rivalry plays out are predominantly economic and technological. This is not to deny the hard power component of the US–China relationship, including China's military build-up, its space ambitions, actions in the East and South China Seas, as well as the risk of military confrontation particularly over Taiwan. However, whereas in the case of the Cold War global competition was primarily military in nature, the main currency of twenty-first-century great power rivalry is economic and technological. It is through primacy in these areas that the alleged superiority of one's model of governance is sought to be proved.

Two sets of considerations stem from this analysis, which condition the global implications of Europe's energy transition. The first revolves around the changing nature of globalization, interdependence and connectivity. A fundamental difference between the Cold War and today's international system is the categorically different nature of economic interdependence. Economic ties between the West and the Soviet Union were down to the bone, as Soviet communism and Western capitalism coexisted in largely non-communicating social and economic spaces. Today, interdependence and connectivity are prime features of our world.

On the one hand, such interdependence increases the potential for friction and disputes. During the Cold War, the segregation between blocs meant that superpower competition and the risk of open conflict played out in a single, dramatic, domain – the nuclear one. There was hardly any political, societal and economic interaction between the Soviet Union and the West, circumscribing the scope of tensions, disagreements and conflicts. By contrast, today's competition is multifaceted: it is about trade, investment, technology, space, energy, public health, information as well as security and defence (Rachman 2021). Competition could spill into conflict in any one of these areas, at the risk of dragging along all others with them.

On the other hand, interdependence mitigates the risk of violent conflict. Precisely because of the existence of multi-faceted interdependences across different sectors, the stakes in a complete breakdown of relations are unbearably high for all. Communication and cooperation in one area can be used at least to temper the risks generated by disagreement and conflict in another. Furthermore, unlike the Cold War, countries are not divided into clear-cut blocs, but oscillate more fluidly between one pole and another. Most players – including the EU – resist the proposition of having to choose between the US and China. All players, including the latter two, acknowledge that the pursuit of global public goods requires international cooperation given global interdependences and the transnational nature of twenty-first-century challenges. In other words, the international system today is far more interconnected and liquid compared to the past.

This said, the tide is clearly turning against interdependence and cooperation, with growing trends towards protectionism, closure and new cold (and hot) wars across different world regions. These trends were reinforced by the pandemic and the Ukraine war, accelerating momentum behind the reshoring and nearshoring of economic activities, also in view of the heightened need for greater diversification and security of supply (Ciravegna and Michailova 2022). In particular, the severe cutback in international business travel during the pandemic and the heightened concern about the West's dependence on China as the world's manufacturing hub, accelerated the drive towards greater diversification within Asia, as well as the nearshoring of production in the Americas and the European neighbourhood (Legrain 2020; O'Neil 2021). The trend towards greater closure is present in China too, where notwithstanding the Belt and Road Initiative (BRI), the 2021–5 Five-Year Plan put a renewed focus on domestic economic activity, better known as the 'dual circulation' approach centred on domestic production for domestic consumption (Tran 2021). Aware of protectionist trends worldwide as well as the crystallizing confrontation with the US, the Chinese regime appears to be hedging its bets, refocusing on its domestic economic activity to sustain growth in anticipation of a potential deceleration of its international economic projection (Hass 2021). Whereas during the Trump administration, Beijing prided itself as a champion of multilateralism and free trade, its successive domestic reorientation, alongside its unfair trade practices and protectionism, certainly do not make China a bulwark of economic interdependence and openness.

Even more dramatic is the turn of the political tide in the US, given the political values it represents. During the Cold War, the US regarded economic integration in the free world as a good to be cherished, and then pursued this ideal globally with the fall of the Iron Curtain. However, beginning with the Trump administration, the US began questioning the belief that economic openness is inherently good, notably for the American people. President Trump made closure, be it in terms of trade or migration, his principal political banner, considering multilateral cooperation, alliances and partnerships either harmful or an unnecessary waste of time for the pursuit of 'America first'. President Biden has explicitly shunned his predecessor's 'America first' agenda, making the rebuilding of international relationships a cornerstone of his foreign policy doctrine (Brands 2021). Yet the US has not rewound back to the past by embracing trade liberalization. It has not re-entered the Trans-Pacific Partnership, nor did it relaunch talks with the EU on a Transatlantic Trade and Investment Partnership. 'Buy American' is here to stay for now, and 'America first' has morphed into 'American manufacturing first' (Nolan 2021). At most, the US and EU have succeeded in removing the most acute economic irritants bedevilling transatlantic relations, including the long-standing dispute between Airbus and Boeing, or that on steel and aluminium tariffs. By pursuing a 'foreign policy for the middle class', Joe Biden has implicitly embraced a soft form of protectionism. This suggests that the turning tide against openness in the US is structural and not merely agency-related and reflective of the proclivities of former President Trump.

Any US president, including one well-disposed towards cooperation, interdependence and partnership like Biden, thus must adapt in order not to lose touch with the electorate (Baer 2021; Drezner 2021).

There is an explicitly Chinese twist to this protectionist story. With the crystallizing US–China rivalry, featuring China's use of economic statecraft to pursue strategic goals as well as a growing Western awareness of the security vulnerabilities underpinning open economic ties, the push towards economic decoupling in the US has grown. The decoupling of the US and Chinese economies is unlikely to manifest itself in all sectors. Doing so would be practically impossible and hugely costly. However, as the weaponization of the economy (Farrell and Newman 2019) becomes a common trait of twenty-first-century great power rivalry, the push towards decoupling is likely to pan out across different critical sectors, beginning with the digital and technological ones and, as discussed below, possibly energy as well.

The second implication of the emerging bipolarity is that Europe, and particularly the EU, is likely to play a fundamentally different role in today's global competition than during the Cold War. In the twentieth century, Europe mattered to the US – and to the Soviet Union – because it was on the proverbial menu. The Cold War began in Europe, slicing the continent into two political halves. Against the backdrop of shared transatlantic values and deep economic, societal and cultural bonds, US presence in and attention to western Europe was premised on the threat posed by the Soviet Union. In the post-Cold War period and the liberal international order that ensued, the EU's actorness gradually came to the fore, often representing the 'soft' complement to US hard power, notably through its integration and cooperation policies in neighbouring regions. US global hegemony and the liberal international order represented the broader global context within which EU enlargement, neighbourhood, development, trade and multilateral policies unfolded.

With the onset of the global power shift and the first signals of multipolarity in the 2010s, European transatlantic anxieties started rising (Tocci and Alcaro 2014). When

Barack Obama's administration announced the US's 'pivot to Asia', alarm bells began ringing in Europe (Rifkind 2011). These reached unprecedented levels when President Trump treated Europeans, and particularly the EU, as public enemy number one, slapping tariffs on the Union, threatening extra-territorial sanctions over the Iran nuclear deal, and questioning the solidity of NATO's Article 5 on collective defence. Under President Biden, that sorry page in the transatlantic relationship has been turned, at least for now. But European anxieties over the US have not dissipated. As the US's strategic reorientation towards China consolidates, many Europeans fret about the transatlantic implications this may have. Nowhere was this clearer than in the European reactions to the US's withdrawal from Afghanistan or the deal over military submarines between Australia, the United Kingdom and the United States (Tocci 2021c). Whereas in the context of the war in Ukraine transatlantic cooperation and coordination were restored to levels almost hitherto unknown, there is full recognition in Europe that had a president of the likes of Donald Trump been in the White House – a future prospect that is far from unlikely – this would have been impossible.

It is certainly true that Europe is – luckily – not going to be the foremost region in which the global competition between the US and China plays out. That unwanted trophy is for the Asia Pacific to bear. However, because of the primarily economic and technological features of twenty-first-century systemic rivalry,[20] the EU is playing a different, and arguably far more important, role than during the Cold War. The EU is often derided for being an economic giant, a political dwarf and a military worm, as Belgian foreign minister Mark Eyskens infamously described it in 1991 (*The Economist* 2017). While on one level that unflattering description continues to apply, on another it is far removed from the realities of today.

Europe's anxieties are in many ways misplaced. If it is true that the US–China strategic rivalry is primarily taking place in the economic and technological domains, then far from being on the menu, the EU sits around the table. The EU's economic weight and competences in these areas place it in a totally different position at the global level. It is a global economic and technological table on which rivalry and competition will be served, including through the possible use of sanctions. This regards not only the US and China, but the EU and China as well. However, given interdependences and the links between the economy and technology on the one hand and the global public goods agenda on the other, cooperation is also being sought. In other words, there is no fine line separating economic and technological competition from the need to cooperate on global public goods. Nowhere is this clearer than in the fields of climate and energy, as discussed below. Given the EU's economic weight and its global leadership in climate and the energy transition, European choices, especially in the areas of natural resources and industry, investments, carbon pricing and finance, will affect deeply both the competitive and the cooperative dynamics within the broader international system.

The geopolitics of transition and great power rivalry

Chapter 3 briefly touched on the fact that a decarbonized world is likely to be more decentralized, with renewable energy sources being more abundant and evenly distributed compared to fossil fuels. This means that markets are likely to become less oligopolistic and more competitive, with renewables also lending themselves to the production of smaller quantities by smaller and more local players (O'Sullivan et al. 2017; Scholten 2018). With most, if not all, countries able to exploit some combination of renewable energies, the climate conversation often features talk about 'prosumers': a growing number of players that will be energy producers and consumers at the same time. Geography will still matter, with different countries having different potential for solar, wind, biomass, hydro, oceanic or geothermal energies, depending on their location and topography. Some will be better endowed than others, with ensuing differences in their abilities to loosen their energy import dependences.

However, all countries will have more choice concerning the balance between domestic production and security of supply versus energy imports and affordability.

A world in which renewables play a prominent role is also more electrified, given the larger portion of final uses that will be converted to electricity in a decarbonized energy system. Electricity is generally traded locally and regionally rather than globally like fossil fuel molecules shipped across the globe. This is because it is much harder and more expensive to transport electricity over long distances, although the development of high-voltage and direct current transmission technology has made long distance electricity transport more feasible and affordable than it once was. This means that a decarbonized world is probably going to be more localized, regionalized and perhaps less globalized (O'Sullivan et al. 2017; Scholten 2018; IRENA 2019a; Bordoff and O'Sullivan 2021). This tends to reinforce existing trends in (de)globalization, notably those towards protectionism, reshoring, nearshoring and regionalization. The IEA (2021c: 282) estimates that international energy trade will fall from $1.5 trillion today to $0.9 trillion in its 2050 net zero scenario, despite the significant growth in the global population and economy expected by then.

A more decentralized and regionalized energy landscape does not entail a depoliticization of energy, however. Precisely because energy sources will be more abundant but also more diffuse, the technologies and systems to capture, store and transport them will become more important in future (Overland 2019). Energy assets will move down the value chain, from commodities to technologies, in wind, solar, batteries, electrolysers, high-voltage transmission, CCS as well as nuclear. This may lead to a growing convergence between energy leverage and power in international relations (Hafner and Wochner 2020). In other words, energy in a decarbonized world may play a far more visible role in global geopolitics, rather than being under the surface and at times a sideshow of great power politics. The US's formidable entrepreneurial spirit; China's growth, its state-driven policies and head start in clean technologies; and Europe's climate leadership, market and regulatory power suggest that all three global players will wield significant energy influence in world affairs.

With energy climbing up the ladder of global players' power resources, the energy transition is likely to become increasingly trapped in competition between them. Climate action is recognized by all as an area where cooperation is essential. As a quintessentially global public good problem, the amount of aggregate effort is essential. Until and unless most of the significant actors cooperate, the goal of remaining within a 1.5°C rise in global average temperatures will not be achieved. Acknowledging this, the US, the EU and China have all declared that climate change is an area of necessary cooperation. The US and China recognized as much despite their growing rivalry (US-DOS 2021b), reaffirming their mutual commitment in COP26 in Glasgow (US-DOS 2021c). The US, though, has clarified also that the US and China are destined to compete across different sectors (Kim 2021), including the economy and technology. While not specifying energy, this inevitably falls into the mix given its tight connection to both. Yet energy and climate are joined at the hip, making separation between them both conceptually and practically impossible. Hence, even if all players firmly believe in and assert the imperative of cooperating on climate change, they will not necessarily succeed in doing so given the mounting competition between them over the economy and technology, and therefore energy too. As ominously put by the Chinese Foreign Minister in response to the US strategy to compete with China in all areas except climate: 'surrounding the oasis [of climate] is a desert ... and the oasis could be desertified very soon' (Reuters 2021a).

The risk of desertification of the climate oasis is two-fold. First, as great power rivalry between the US and China intensifies, possibly spilling into open confrontation in Taiwan or the South China Sea, it is difficult to imagine that climate could be shielded as a protected space of enduring cooperation. Whereas some may argue that the US–China competition could ratchet up climate action (Erickson and Collins 2021), it is more likely that were competition to tip into confrontation in one policy field, it would pollute all others too, climate included. Second, to the extent that great power rivalry is fanning the flames of closure, protectionism, reshoring and decoupling, this could feed energy nationalism in ways that would hamper the prospects of climate cooperation and the energy transition. The US's tariffs on solar panel imports are a case in point. Less dramatic, but in the same vein, is the US move on domestic tax breaks or direct payments that would apply to specific industries, including wind, only provided domestic content requirements are met. Whereas the former measure was clearly targeted against China, the latter would affect others too, Europeans included. A decarbonizing and decarbonized energy system is one which is inextricably linked to the economy, technology and security. Believing in a dematerialized energy transition that can be detached from the conflict and tension of the material world is a pipedream.

The risk of global rivalry over energy and climate, as well as the drive towards nationalism and protectionism, must be read also against the backdrop of the new dependences inherent in the energy transition. Whereas the transition away from fossil fuels implies a reduction of energy dependences, these will not cease altogether but rather be replaced by other decarbonized dependences. New dependences are emerging, notably over critical minerals and green tech supply chains more broadly (Hache 2018). Here too, energy influence and power in international relations increasingly converge.

Whereas minerals needed to produce renewable technologies are distributed across different regions, China is well ahead in the critical minerals race (Stratfor 2018). There are around thirty raw materials considered 'critical' in a digital and decarbonized economy, due to their scarcity, lack of substitutes and strategic relevance to supply chains (Szczepański 2020). Rare earths, key to the production of electric vehicles, wind turbines and many other low-carbon technological solutions, are mainly produced in China, making the mining and refining of rare earths a de facto Chinese monopoly. Other critical minerals like lithium or cobalt, essential for storage, are mined predominantly by

Chinese companies (Neill and Speed 2012). Of the five companies that account for approximately 90 per cent of global lithium production, for instance, three are Chinese or feature significant Chinese capital. Chinese companies have also established a near monopoly over cobalt in the Democratic Republic of Congo (DRC), where almost half of the global supply is located (Gulley et al. 2019). China's presence in DRC is certainly not uncontested domestically but remains dominant nonetheless (al-Jazeera 2021).

Coupled to this, China has an edge in the manufacturing and deployment of solar, wind and storage technologies, it boasts the largest market in electric vehicles, and is fast developing the equipment and technologies to manage smart grids. In solar energy in particular, China accounts for the manufacturing of two-thirds of the world's polysilicon and as much as 90 per cent of the semiconductor wafers necessary for solar power cells (Bordoff and O'Sullivan 2021). As figure 3 highlights starkly, solar manufacturing is essentially in Chinese hands.

Even if the US, Europe or other players massively launch and foster green industrial capacities, it will take time for these to become competitive with China and to fill the supply gap if China were to be hypothetically side-stepped. A decarbonized Europe without China would be prohibitively costly, and probably unfeasible. As provocatively put by Allison (2021): the green future may in fact be red.

This does not mean that China has won the energy transition race or that it can leverage or weaponize its energy transition assets to extract political or strategic gains. China is the world's greatest emitter because of its enduring dependence on coal. The high carbon intensity of China's economy means that decarbonization will come at a high cost. Furthermore, China's coal plants are relatively young, with an average age of less than fifteen years (Cui et al. 2021; IEA 2021c). With the average life of a coal plant being over forty years, the depreciation that would come with a phase-out of coal in the next decade is huge. All this while the regime knows it cannot afford to slow down growth, lest this ignite societal grievances and political disaffection and

(a)

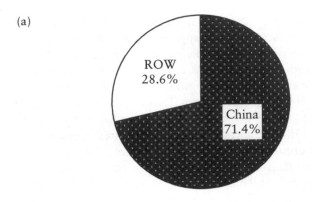

(b)

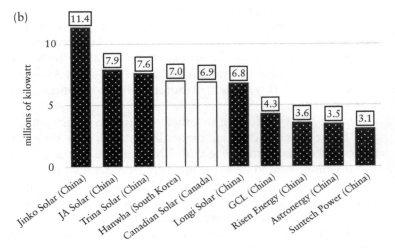

Figure 3: World solar PV manufacturing (2018): (a) Market share – China vs ROW (2018); (b) Top 10 manufacturers (2018)

Sources: Graphs from Tocci (2021a). Data from BNEF, IHSMarkit, IAI. Reprinted with permission.

protest. Moreover, employment in the coal sector represents an important social safety net in some provinces such as Shanxi and Inner Mongolia, making their closure not merely an energy or economic issue, but a social and political one too. No surprise that, whereas President Xi pledged to end international coal projects in 2021, divesting from coal domestically is not on the near-term horizon, certainly not as China seeks to propel its post-pandemic growth and grapples with energy supply shortages and high prices. No wonder too that China reacted negatively to the notion of carbon tariffs that would hamper considerably the competitiveness of its carbon-intensive exports, arguing that such levies are not necessary given the commitments made under the Paris Agreement (Sapir and Horn 2020). In short, the energy transition entails heavy costs, and China will bear a considerable share of them.

China's edge in green technologies must also be unpacked. In wind energy, for instance, if one discounts installations in China, European manufacturers are dominant globally (Jaganmohan 2021). Figure 4 highlights Europe's significant global share in wind manufacturing. On innovation, while it is true that most renewable technology patents are Chinese, the majority of these are filed in China itself (Lam et al. 2017). Furthermore, there is much more to transition than renewables as we know them today, and innovation continues to be fostered primarily in the West. One only needs to think of the potential of breakthrough technologies like nuclear fusion to highlight this point.

On critical minerals too, China's dominance must be contextualized, and we cannot simply assume that China will exert leverage in a decarbonized energy system, like OPEC did in the fossil fuel world (Overland 2019; Krane and Idel 2021). On the one hand, whereas it is true that China has already attempted to weaponize its exports of critical minerals, for instance to Japan in 2010, its ability to do so in a sustained manner is limited. China's ability to leverage systematically its control of critical minerals is limited by the WTO, where the EU has already demonstrated its ability to induce China to give up its export restrictions in the past. On the other hand, whereas the security of oil or gas supplies revolves around flows, that of critical minerals is a matter of stocks. If gas supplies through Ukraine were cut for two weeks in winter, eastern Europe would literally freeze. If instead the supply of cobalt from DRC were interrupted for two weeks as a result of an attempted act of trade

(a)

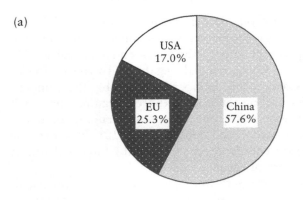

(b)

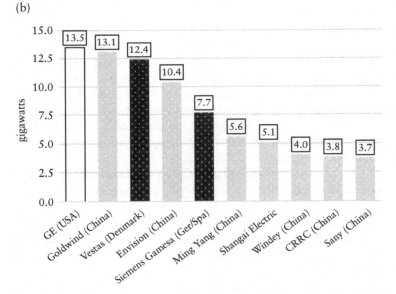

Figure 4: Global wind turbine manufacturing (2020): (a) Global market share (2020); (b) Top 10 manufacturers (2020)
Sources: BNEF, GWEC.

coercion by China, the world would probably not notice at all. This is not to deny the potential vulnerabilities of green supply chains, notably critical minerals. These largely revolve around problems of social and environmental sustainability. China has had a head start in the mining and production of critical minerals not just because of its far-sightedness but also because of its less stringent or non-existent labour and environmental norms in this area. This is unlikely to change any time soon, making the global governance of this sector of crucial importance. However, it is wrong to project the current concentration of control over critical minerals into the future and derive from it the conclusion of an unsolvable geopolitical conundrum.

In fact, Chinese dominance may not be sustained in future, as mining will proceed in different parts of the world and as China's own demand for critical minerals will increase in order to meet its clean energy targets. Furthermore, technological developments, including on recycling, could make the need for such raw materials less critical in future. In other words, China today plays a dominant role in the transition landscape, although this is not set in stone. The snag is that we are at a minute to midnight on climate change, as emphatically put by British Prime Minister Boris Johnson in his opening of COP26 in Glasgow (Rowlatt 2021). The timeframe to reach climate neutrality is very short indeed. Given the acceleration of this policy-driven energy transition necessary to limit global warming, China will inevitably be a crucial piece of the puzzle in the race to net zero.

European strategic autonomy and the energy transition

Where does Europe's energy transition fit into this complex global picture? The EU has set for itself the goal of strategic autonomy (Tocci 2021a). The term was first coined in the context of security and defence, notably in the EU Global Strategy (EEAS 2016). It gradually migrated across different policy sectors, including the economy, finance, digital as well as energy. This expanded understanding of autonomy was adopted by President Ursula von der Leyen's 'geopolitical' Commission (Bayer 2019).

The pursuit of European strategic autonomy is intertwined with the energy transition. Given that the EU is a consumer rather than a producer of fossil fuels, decarbonization and the development of European green capacities offer the opportunity to enhance European strategic autonomy, or, put in energy lexicon, European energy security. The reduced fossil fuel dependence on Russia, first and foremost, would strengthen enormously European autonomy and resilience.

However, the EU is cognizant of the new interdependences embedded in green supply chains, from critical minerals to green technologies, and how these could also be weaponized, exposing Europeans to new security risks. A few figures highlight the risk. The EU today produces only 1 per cent of the raw materials to produce lithium-ion batteries, 8 per cent of the processed materials and 0 per cent of overall assembly (JRC and European Commission 2020). In a net zero EU by 2050, European imports of lithium are expected to increase sixty-fold, those of cobalt fifteen-fold and those of rare earths ten-fold to meet targets on renewables, storage and electric vehicles (Vanheukelen 2021). The EU's drive to avoid falling prey to these new energy dependences is reinforced by the desire to deliver on its promise to decouple growth from emissions, demonstrating that the European Green Deal is also the recipe for European prosperity. If a green Europe is one that delivers on growth and jobs, green capacities would need to be developed at home rather than simply be imported from abroad. In this respect, the European debate reflects that in the US and President Biden's Green New Deal. Driven by the need to secure supplies and strengthen energy resilience, as well as to mobilize green growth, the EU is establishing institutional, regulatory and funding mechanisms to foster European green industries and technologies.

An integral part of the EU's strategy in this respect is that of establishing industrial alliances. Focused on critical sectors for a more resilient, autonomous and prosperous Union, these bring together European public, private and civil society players along the value chain. Alliances are not directly funded by the EU, but projects developed within them are eligible to become 'important projects of common European interest', that are exempt from state-aid rules. Interestingly, of the six industrial alliances that have been launched, four – raw materials, batteries, clean hydrogen and circular plastics – are directly related to the energy transition, signalling its prominence in debates surrounding European prosperity, resilience and autonomy.

The European Battery Alliance brings together hundreds of public and private players with a view to creating an industrial ecosystem conducive to the development of European storage capacities. Within it, the Commission has funded several projects of common interest, such as the 2021 €2.9 billion battery innovation project (EBA 2021). The importance of storage capacities is growing, especially as Europe affirms itself as an important player in renewables, particularly wind. In 2020, despite the pandemic, European investments in renewables soared, with €82 billion spent on renewable capacity investments (Ren21 2021), of which over half was on the construction of new wind farms (WindEurope 2020). In its strategy for offshore renewables, the Commission foresees a thirty-fold increase in offshore capacity by 2050, worth €800 billion of investments (European Commission 2020c).

Similar, albeit more nascent, is the clean hydrogen alliance. Given the awareness that not all sectors are prone to electrification, the EU's green agenda has placed much hope on clean hydrogen molecules to fill the gap. Indeed, it is difficult to imagine the decarbonization of hard-to-abate industries, to name the most obvious example, without clean hydrogen playing an important part. Here too, the aim is to bring together players along the hydrogen value chain, from renewables and low-carbon hydrogen production, via transmission, to mobility, industry, energy and heating applications. Also, in this case the idea is to promote projects of common interest. These initiatives, alongside different forms of top-down market creation, work simultaneously towards the goals of decarbonization and strategic autonomy.

The raw materials alliance is another interesting case. The EU has long identified the strategic relevance of critical raw materials, developing its first list in 2008 and progressively updating it since then. More recently, it has established an industrial alliance born out of the recognition of Europe's dependence on raw materials. The alliance aims at building European resilience along the rare earth and magnet value chains, supporting sustainable and environmentally friendly domestic mining and processing, while reducing dependence on such raw materials through the development of recycling. China, as the largest supplier of critical raw minerals to the EU, accounting for just under half of them, is not mentioned in the alliance's official documents. But it is precisely its absence that speaks louder than words. The alliance includes well over 100 companies, institutions, associations and universities. They are mainly European but stretch beyond the EU, from Australia to India, Malaysia to Brazil. China does not feature on the list.

Working in this direction is crucial. Addressing fragmentation, building partnerships and inducing investment through institutional, regulatory and financial measures is critical for the development of European green capacities. While not hampering intra-European competition, the EU must enable itself to compete internationally by fostering green industries. The approach taken on batteries, hydrogen and raw materials indicates a healthy balance between the need to protect internal openness and competition while promoting resilience, international standards, competitiveness and a level playing field.

It will not be easy. Creating industries is costly and time consuming. Beyond filling gaps in skills and technologies, the EU must develop and enforce norms and standards, whilst overcoming the political opposition that may arise. Nowhere is this clearer than in the area of critical minerals, in which lengthy industrial processes, the imperative of respecting environmental standards, the technological progress that still needs to be made,[21] and the possibility of social and political resistance conspire against delivering on the EU's objectives. In fact, given the high environmental impact of mining, the coming years may witness a growing tension between the climate and environmental dimensions of the European Green Deal, as well as between the latter and the goal of strategic autonomy. Whereas the EU's climate and strategic autonomy goals would push for the development of critical mineral mining, including in Europe, environmental and social concerns may warrant against it.

Building internal capacities is necessary to pursue European decarbonization and strategic autonomy. But it is neither easy nor sufficient. Promoting 'open, transparent, well-regulated, liquid, and rule-based global markets ensuring a diversity of suppliers and sources' is equally important (European Council 2021a). The Union strives for 'strategic autonomy while preserving an open economy' (European Council 2020). Particularly in the context of trade, industrial and digital policies, Executive Vice President of the Commission Margrethe Vestager in fact coined the term 'open strategic autonomy', signalling that European autonomy is not tantamount to independence or autarky. In so far as multi-lateralism and interdependence are staples of the European diet, European autonomy can only be understood as 'open'. A closed and protectionist EU would run counter to the very DNA of the European integration project.

On the surface, this is a no-brainer. EU strategic autonomy does not mean that everything has to be produced in Europe: not only is this unnecessary but it would also be unbearably expensive. It would be unwise for the EU to try to outcompete China in labour-intensive manufactured goods such as solar panels, for instance. Likewise, in the digital sphere, it is difficult – and unnecessary – to imagine that the European alliance on processors and semiconductors would be able to outcompete South Korea or Taiwan. By contrast, boosting existing EU comparative advantages in high added-value sectors would make more strategic sense, from wind turbines to electrolysers. The EU is already a global leader in these areas, and it can consolidate its gains, strengthening further its resilience.

On closer inspection, however, difficult trade-offs loom on the horizon. In the strategic autonomy debate, particularly as it pertains to industry, including energy, hardening European views on China have entered the scene. Europeans disagree with those Americans who claim that it is possible or even desirable to decouple economically from China. However, China is no longer viewed just as a strategic challenge in East

Asia, while being seen as a rather benign economic force elsewhere. The strategic edge of China's BRI and economic statecraft has become clear to Europeans since the mid-2010s; and Beijing's propensity to use technology for repression and propaganda has been an eye-opener, particularly since the pandemic. Today, Europeans are far more clear-eyed about the security and political risks entailed in China's espionage, forced tech transfers, strategic commercial interactions and asymmetric agreements, and far more willing to pursue regulatory and political measures to hedge against these. The EU has already tightened its screws on anti-dumping and investment screening, as well as seeking to regulate the access of state-subsidized enterprises (European Commission 2020d). Its anti-coercion package aims at deterring third countries like China from restricting or threatening to restrict trade and investment to leverage change in EU policies in areas like climate, taxation and food safety. Europeans are also far warier of the tightening political grip of the Chinese regime on Hong Kong, its growing bullishness towards Taiwan, its political backing of Russia in its invasion of Ukraine, as well as human rights violations, beginning with forced labour in Xinjiang.

This raises two dilemmas, political and economic. Politically, the cooling of EU–China relations forces Europe to choose between two values. In the European foreign policy debate, the juxtaposition is often made between values and interests, where European players, while rhetorically supporting the former often cynically pursue the latter, particularly vis-à-vis third countries where strategic interests are at stake (Tocci et al. 2008). For years, that dilemma has put the spanner in the works of the EU's self-depiction as a normative power in the world (Manners 2002). But the nexus linking the EU, China and the energy transition highlights a new, and arguably even harder, set of choices ahead. Commission President von der Leyen (European Commission 2021i) declared: 'Doing business around the world is good, global trade around the world is good and necessary, but can never be done at the expense of people's freedom and dignity ... We will propose a ban on products made with forced labour. Human rights are not for sale – at any price'. This relates to the broader drive within the EU to enshrine due diligence in law, making companies accountable if their suppliers breach labour and climate standards and human rights in general. Brussels has already made its first steps to introduce limited import bans, including on products from areas at risk of deforestation, as well as proposing legislation on batteries that would require companies to assess the human rights risks entailed in their supply chains. While no country is named, China immediately springs to mind, notably its repression of Uighurs in Xinjiang. In a similar vein, the US has moved to ban imports from China believed to be produced by child or forced labour (Kaplan et al. 2021).

The snag is that China is not only the global leader on solar energy, but much of the world's polysilicon used to produce photovoltaic cells comes from Xinjiang. With a very tight supply of polysilicon, it is difficult to imagine that all European (and indeed American) companies in the solar business will be able to turn down supplies from Xinjiang (Murtaugh 2021). Some may argue that protecting human rights comes at a price, and if standing against forced labour means more expensive solar energy by detouring China then so be it. The EU plans to invest significantly in its solar industry, and nearly all elements of the solar supply chain, including solar-grade polysilicon, can be produced in Europe, albeit at higher costs (Hernández-Morales et al. 2021). But arguably it is not just a matter of economic cost, but actual feasibility within a given timeframe: it is difficult to imagine that the five-fold increase in European renewables to ensure a 55 per cent reduction of emissions by 2030, on the way to net neutrality by 2050, can be met without imports from China, including Xinjiang in the years ahead. Rather than values versus interests, the juxtaposition here is between two sets of values: human rights and net zero greenhouse gas emissions.

Economically, the complex relationship with China places the EU in the uncomfortable position of toying with protectionism. Ideally and in line with European 'open strategic autonomy', the EU would instil greater openness and transparency in China's political economy. After seven years of negotiations, the 2020 EU–China Comprehensive Agreement on Investment (CAI) was intended to achieve a levelling of the playing field, market access, limits to forced technology transfers and, in principle, labour standards. The ratification of the agreement has been suspended in response to China's sanctioning of a group of European parliamentarians and researchers. But regardless of whether ratification will proceed eventually or not, the question looms: if China continues practising protectionism and unfair trade, and the EU must react to ensure the resilience of its industry and economy, will a levelling of the playing field mean the EU's move towards protectionism rather than China's shift towards liberalization?

On the flip side of protectionism is the risk that systemic rivalry with China and the drive to develop European green capacities in response to it could hamper the EU's sustainable development goals in the Global South in the name of free trade. In light of the growing importance of critical raw materials in the energy transition, resource-rich countries in the Global South understandably aim to develop the more profitable components of green industries further down the value chain. Countries like DRC and Chile have already signalled their intent to build their domestic battery industries given their abundance of lithium and cobalt respectively. However, in the name of free trade, the EU opposes these suppliers' export restrictions of raw materials. In fairness, the EU has been consistent in such opposition across all sectors, geographies and time periods. This was a major irritant in EU–Chile negotiations to revise the free trade agreement between them. However, today, alongside the EU's general rejection of export restrictions is the drive for European strategic autonomy in the energy transition, which is linked to the strategic rivalry with China. The EU opposes the restrictions of exports of raw materials also because of its own interest in developing such industries rather than relying predominantly on China. But whereas when viewed from the perspective of autonomy and free trade the EU rightly opposes export restrictions that countries like DRC or Chile might want to enact to support the development of their own green capacities, seen from a sustainable development angle in the Global South the EU's position does not hold water. In fact, were these countries not to foster their own industries up the green value chain, their own development would be hampered, and new forms of extractionism would be promoted and legitimized.

In practice, these normative dilemmas need not be so black and white. The EU can foster its autonomy and the green agenda without completely turning a blind eye to forced labour, paying prohibitively high costs, hampering sustainable development, or giving in to protectionism. It can square these circles by diversifying and greening its partnerships with countries and regions around the world. One only needs to think of how, in the pursuit of net zero emissions, the EU is factoring in relations with North Africa and Ukraine, for instance through the 2×40 GW initiative within its hydrogen strategy, or how it is pursuing a clean energy and climate partnership with India. Likewise, in the development of critical raw materials, the EU is eyeing partnerships with developed mining countries like Canada and Australia, as well as developing and emerging countries in Africa, Asia and Latin America. The partnership between the EU, the US, the UK and South Africa, launched in the context of COP26, is another case in point. The same ethos permeates the EU's Global Gateway initiative, the EU's connectivity strategy that seeks to provide a green response to China's BRI (Tanchum and Murphy 2021). While representing to an extent a repackaging of existing funds and programmes, the Global Gateway, with its €300 billion in funds to be spent by 2027, represents an important landmark bringing together and communicating effectively the EU's infrastructural and global drive on connectivity embedded in European green norms. Regionalizing and even globalizing the transition beyond the EU's borders will be essential to reconcile autonomy, norms and costs, or, put differently, the energy security–decarbonization–affordability trilemma. Working with surrounding regions will be key to protecting interests and values, containing economic costs and accelerating decarbonization. Finding common cause with the United States is just as important.

A transatlantic green cause

While there is no silver bullet to pursue concomitantly the EU's climate neutrality, strategic autonomy, prosperity and values, the best available recipe includes a stronger transatlantic green partnership. Working together with the US on the energy transition means building critical mass on energy and climate policies that others, notably China, would find difficult to resist. It also means cooperating on the development of green industrial capacities and resilient green supply chains, striking a balance to ensure security without skyrocketing costs.

However, climate has often been an area of transatlantic divergence, with the EU overtaking and now being well ahead of the US in its climate ambitions. Up until the 1990s, climate policy was not politically divisive in the US, notably between Democrats and Republicans. In fact, it was under the Republican presidency of Ronald Reagan that the US led the global fight to address stratospheric ozone depletion, more commonly known as the ozone hole, which eventually crystallized in the 1987 Montreal Protocol on Substances that Deplete the Ozone Layer. The US was also pivotal in the early scientific research into climate change and the ensuing establishment of the UN's IPCC in 1988. However, in the years that followed, the tune changed drastically in the US. Over the course of the 1990s, opposition to climate action became widespread in the US, particularly within the Republican Party. A 2019 Pew Research Center survey found that whereas 30 per cent of US citizens believed that climate policy would hurt the economy, that figure was over 60 per cent amongst Republicans (Funk and Hefferson 2019). Climate action became associated with big government and juxtaposed to the free market, being characterized as an impediment to growth, prosperity and jobs. The political power wielded by fossil fuel industries, the ingrained resistance to regulation and the role of the state in the economy, and the general attachment to an 'American way of life' revolving around high consumption and ever-growing material prosperity, all contribute to explaining the powerful forces against decarbonization in the country (Collomb 2014). Europe, by contrast, has never witnessed such climate polarization. Until recently, climate change was not even viewed as a particularly salient political subject, with denialist views being limited to the political fringes as discussed in chapter 2. Generally, there has been broad agreement in Europe concerning the fact that decarbonization is a goal worth pursuing. This has enabled EU institutions to surpass the US in climate leadership since the 2000s and to raise the level of climate ambition over the years, with the European Green Deal representing the clearest and most concrete example.

Transatlantic divergence on climate and energy policies on some occasions contributed to a global climate impasse. At the 2009 COP15 in Copenhagen, a wide transatlantic gap emerged over the role of the 1997 Kyoto Protocol in reaching a new agreement. Whereas the EU remained anchored to the imperative of building on the Protocol and its binding targets, the Obama administration – notwithstanding its greater commitment to climate compared to its predecessor – was unwilling to reconcile itself with Kyoto, from which George W. Bush had explicitly taken distance in 2001. Without an agreement between developed countries – notably the EU and the US – bridging the divide with developing countries became impossible. Failure to reach a global agreement was the result (Rosen 2015). Other COPs, notably COP21 in Paris, were successful. However, the negotiations leading to the 2015 Paris Agreement saw the EU and the US operating on different, although in this case complementary, levels. The Paris Agreement rested on two pillars. First, the European level of climate ambition[22] and its ability to build a 'high ambition coalition' with a group of seventy-nine African, Caribbean and Pacific countries. Second, an agreement between the US and China.[23] Both pillars were essential and complementary, but they did not imply the transatlantic partners working hand in hand (Davis Cross 2018).

The transatlantic gap widened more than ever during Donald Trump's administration. Not only did President Trump withdraw from the Paris Agreement, but his systematic attack on multilateralism, diplomacy and the EU precipitated the transatlantic relationship to reach an all-time low. Transatlantic tensions are not new, with relations historically witnessing ebbs and flows (Cowles and Egan 2012). Over the decades, transatlantic ties have periodically frayed over trade, foreign policy, security and defence. There have been banana wars[24] and deep frustrations over European passivity in the Balkan wars, cultural wars over genetically modified organisms, and profound European anger over US policy in the Middle East. There have been times in which millions of Europeans took to the streets to protest against the US invasion of Iraq, whilst Americans renamed French fries as 'freedom fries'. Periodically we have been through tough transatlantic times. However, in the past, frictions and disagreements played out within the borders of a commonly recognized transatlantic family.

This is what changed under President Trump. For the first time, the incumbent of the White House explicitly challenged the shared values that form the bedrock of the transatlantic bond, questioned the US's commitment to NATO's collective defence, and treated European leaders as adversaries rather than friends and allies. The very existence of a transatlantic family, premised on shared political values as well as strong historical, cultural, social, security and economic ties, was in doubt. This, more than policy disagreements over steel and aluminium tariffs, the Iran nuclear deal, or indeed climate change, made transatlantic cooperation next to impossible.

Under President Biden, the EU and the US have been eager to turn a new transatlantic page. As always, there is plenty of debate and even disagreement across the Atlantic. Europeans criticized the US's shambolic withdrawal from Afghanistan, anxious that this might signal a broader US withdrawal from the world. They have stood alongside France, lambasting the agreement between Australia, the UK and the US as evidence that Washington's instinct to put 'America first' while neglecting allies was alive and kicking. The EU and the

US take different views of data privacy and the regulation of big tech. While increasingly convergent, they also nuance differently their approaches towards China. The EU, while highlighting its systemic rivalry with China and becoming increasingly aware of the security risks of China's growing economic and technological penetration in Europe, rejects the notion of a decoupling from the Chinese economy. The economic relationship with China is viewed through the lens both of competition and of cooperation in Europe.

Transatlantic differences aside, on security, in light of Russia's invasion of Ukraine, unity, coordination and cooperation between Europe and the US have been remarkably strong, both in defence terms in the context of NATO, and as regards diplomacy and the imposition of severe sanctions on Moscow. Faced with the dramatic threat posed by Putin's Russia, transatlantic security ties have never been so strong.

On corporate taxation, Europeans have long complained about the fact that tech giants such as Alphabet, Amazon, Apple or Meta pay little tax in countries from which their profits also originate. Some of them – France, Italy, Spain, but also the UK – consequently adopted digital taxes, which according to the US discriminated against American companies. First emerging during the Obama presidency, the issue escalated under Trump, who threatened EU countries with steep tariff rises. The Biden administration adopted a different approach, promoting a global minimum corporate tax, which should limit the ability of multinationals to exploit generous tax regimes and ensure they pay taxes where they generate revenues. On the grounds of a transatlantic agreement, the global corporate tax was endorsed by the G20 and the OECD. This has not only removed a major transatlantic irritant but also implied a first small step towards redressing the major injustices and disparities generated by the age of hyper-liberalization.

On trade, while the Transatlantic Trade and Investment Partnership remains shelved, the EU and the US have removed the most significant trade irritants bedevilling their relationship. They ironed out the Airbus–Boeing dispute, in which the EU and the US slapped billions' worth of tariffs on one another after the WTO ruled that aerospace companies Airbus and Boing benefitted from illegal state aid. They reached an agreement on steel and aluminium, addressing the problem of steel overcapacity, for which China is mostly responsible. Interestingly this agreement, rather than limiting itself to the negative goal of removing trade barriers across the Atlantic, has pursued the positive objective of promoting transatlantic sustainable steel and aluminium industries. This model could and should be replicated in other carbon-intensive industries. Seeing the scope for deeper and more systematic transatlantic cooperation in the areas of trade, economy and technology, the EU and the US set up a Trade and Technology Council in 2021.

Climate is another area that has risen up the transatlantic agenda. With Biden's return to the Paris Agreement and the US's net zero pledge, climate policy and the energy transition became a top priority for transatlantic cooperation. On some issues, important steps forward were made. The above-mentioned steel and aluminium agreement is important both in itself and because it opens the way for transatlantic cooperation on other industries whose carbon emissions are hard to abate. Even more immediately consequential to the fight against climate change is the global methane initiative, spearheaded by a transatlantic agreement. The initiative, aimed at reducing fugitive methane emissions by at least 30 per cent by 2030 compared to 2020 levels, became one of the few concrete successes of COP26.

While very different, both these agreements suggest that when the EU and the US find common cause, they can generate a critical mass that becomes difficult for others to ignore. The steel and aluminium agreement will have severe implications for China, which produces around half the world's steel through carbon-intensive coal-fired blast furnaces. However, this is precisely what makes the transatlantic agreement so important, given that it could spur a revisiting of sustainability standards within the Chinese steel industry to retain its global market share. A 30 per cent reduction of methane emissions may be seen as a low-hanging fruit in the race to net zero, given the relatively contained costs it would entail and its high short-term impact.[25] Not all countries subscribed, with Australia, China, India, Russia and Turkey being the most visible absentees. But over 100 countries did join, representing around 70 per cent of the global economy. Cutting methane emissions is easier and faster than reducing CO_2 in the atmosphere. However, given the potent, albeit more short term, effect of methane on global warming, achieving this quick win is essential in the race to net zero. All this indicates the climate influence the EU and the US can wield when they work together.

By contrast, when the EU and the US fail to agree, multi-lateral governance is weakened. In contrast to the global methane pledge, the inability of the EU and the US to move forward together on the phase-out of coal played no small role in dampening the level of ambition of the Global Coal to Clean Power Statement in COP26. Whereas the Biden administration pledged coal-free electricity by 2035, the US did not sign the Global Coal to Clean Power Statement. Both time-contingent and structural reasons prevented the US from taking a bold step on coal. The climate negotiations in Glasgow were taking place at the time when the Biden admin-istration was seeking approval in Congress of its trademark $3.5 billion infrastructure bill. Committing to a phase-out of coal would have complicated the way towards approval, galvanizing opposition amongst coal-rich regions like West Virginia. Moreover, the administration faced broader consti-tutional constraints, given that the federal level cannot make commitments that fall under state competences in the US. Forty-four countries signed the statement, including fifteen EU Member States (Hook et al. 2021). Indonesia, South Korea, Chile and Ukraine joined the statement too, making it an important first step towards ending the use of coal. However, major players such as China, India and Turkey did not sign, reducing significantly the level of ambition of the initiative. In the rather bland joint US–China statement at COP26, reference was made to the less ambitious 'phasing down' rather than 'phasing out' of coal. Precisely this weaker wording eventually won the day in Glasgow, making it into the final communiqué. The inability of the EU and the US to find strong common cause was not the only reason under-pinning this glass half-empty outcome on coal at COP26, but it certainly contributed to it.

A transatlantic green partnership is easier said than done. Interestingly, younger Europeans are well aware of the high stakes involved and of the difficulty of rising to the transatlantic challenge. While viewing transatlantic climate cooperation as the most pressing priority, they are concerned about the wide gap that separates the two sides of the Atlantic. This is what they express greatest concern about (Stokes 2021).

Transatlantic cooperation on climate and energy is both essential yet highly complex in three main areas. First is the creation of resilient green supply chains. I described above the EU's efforts in this respect, notably its industrial alliances for sustainable value chains in areas like critical minerals, batteries and hydrogen. Seeking to develop such capacities only internally in the EU, however, is either impossible – in the case of raw materials – or prohibitively costly – on almost everything else. European strategic autonomy does not and cannot mean European autarky. The same applies to the US.

While unable to detour global cooperation, first and foremost with China, working across the Atlantic to share best practices and build green industrial ecosystems would go far in making climate neutrality more within reach and developing resilient green supply chains during the journey. It also represents a critical step in supporting other major players like India to reach their net zero goal. European and American investments in renewable capacities in countries like India are crucial. The European Commission has rightly proposed a transatlantic green tech alliance in areas like renewables, storage, batteries, hydrogen and carbon capture utilization and storage (European Commission 2020e). Part of this effort should also feature transatlantic cooperation on cyber security, notably on recognizing potential risks to critical links and mitigating these through coordinated responses (Morningstar et al. 2020). The EU–US Energy Council as well as the Trade and Technology Council are the institutional venues within which such cooperation should be pursued, both bilaterally and in reference to cooperation in and with third countries. In fact, the work strands of these two institutional bodies are so deeply intertwined in the context of the twin challenges of energy security and the energy transition that a merger between them could be considered.

A second, even more complex, area of transatlantic cooperation is carbon pricing. Here the transatlantic gap is much wider (Tooze 2021). Whereas the EU's ETS has been in place since 2005 and represents the most advanced and ambitious cap-and-trade system worldwide, the US does not have a domestic carbon market at federal level and is unlikely to develop one anytime soon. The idea of carbon pricing has been debated since the 1970s in the US, being viewed as a recipe to reconcile environmentalism and the market. However, all attempts to introduce carbon pricing, including by the Clinton and Obama administrations, have failed. Opposition to carbon pricing comes from both ends of the political spectrum. Whereas Republican opposition is related to the costs this would impose on industry, the left of the Democrat Party shies away from carbon pricing, criticizing it as a neoliberal market-based solution. The result is impasse. Administrations that have taken climate more seriously – the Obama and Biden administrations – have concentrated their policy efforts on funding, notably for infrastructure, and regulation, rather than on pricing and taxation, given the political economy complexities entailed by advancing on the latter.

Given that the EU and the US stand at opposite ends of the carbon pricing spectrum, the EU's proposal for a Carbon Border Adjustment Mechanism has raised eyebrows in Washington. Initially the US Congress had expressed interest in the idea (Shouse 2021), as it chimed with the general appeal that protectionist measures have in the country. A carbon border tax against Chinese or Russian carbon-intensive imports was an attractive political and strategic proposition on both sides of the aisle. However, unlike the EU, which has a domestic carbon market, the US does not. Were the US to proceed on a carbon tariff, it would probably be challenged successfully at the WTO. As it became clear that a US CBAM would be highly improbable, the US turned against the EU's proposal, adding a new layer of transatlantic friction. US Climate Envoy John Kerry defined a carbon border levy as a 'last resort' which could seriously imperil trade (Fleming and Giles 2021). While not being amongst the most affected countries, CBAM could in fact reduce the competitiveness of US coal, gas and manufactured product exports to the EU (Kleimann and Eacho 2021). If the US could not impose such a levy, Washington became uncomfortable with the proposition of the EU doing so.

The transatlantic gap on carbon pricing is a problem. A European CBAM will be successful if it drives other countries towards decarbonization, eventually eliminating the need for CBAM altogether, as discussed in chapter 3. Were this not to happen, CBAM, while being a necessary complement to an expanded ETS to avoid carbon leakage and a hollowing out of European industry, could also hamper trade, fuel protectionism, harm development as well as aggravate geopolitical tensions. Limiting or avoiding these negative side-effects depends on the emergence of a critical mass of countries developing forms of carbon pricing, and eventually joining forces in the establishment of climate clubs, as notoriously argued by Nobel laureate William Nordhaus (2015).

The good news is that many countries are slowly but surely moving in this direction, with China being the most significant addition. The Chinese carbon market is still embryonic, covering only a third of emissions and featuring extremely low prices. Furthermore, with most companies being state-owned, they often lack the profit-maximizing logic that underpins the functioning of a carbon market. But the establishment of China's carbon market is nonetheless a beginning. Countries, like Australia, Canada, New Zealand, Mexico, South Korea and Kazakhstan, already have established cap-and-trade schemes, whereas a host of others, from Chile and Colombia to Pakistan and Indonesia, are following the same path.

The absence of the US at federal level represents a major drawback, however.[26] Given its size and weight in terms of the economy and emissions, a global critical mass underpinning a climate club might be difficult to achieve without the US. A transatlantic climate club, however complex, should remain high on the list of transatlantic ambitions (Tagliapietra and Wolff 2021). This is not only because its absence may inadvertently lead to a climate trade war between two major economic powers, both of which are committed to decarbonization (Tooze 2021); it is also because with the EU and the US collectively representing around 40 per cent of global GDP and 30 per cent of global imports, the magnetic pull of a transatlantic climate club would become almost impossible to resist by others. This is true especially if other OECD countries with carbon pricing and with trade agreements or negotiations ongoing with the EU were to be included, including Canada, Chile, Iceland, Japan, Norway, South Korea, Switzerland, the UK and New Zealand.

Today we are far from getting anywhere close to a transatlantic climate club. There are major structural, and largely political economy drawbacks to the feasibility of a cap-and-trade or carbon tax system in the US (Cullenward and Victor 2020). Furthermore, the technical difficulties of establishing climate clubs between countries using different forms of explicit and implicit carbon pricing should not be underestimated. COP26 in Glasgow tackled many questions, from climate targets, to coal, methane and climate finance. Carbon pricing was the great absentee, which was only addressed indirectly through negotiations on Article 6 of the Paris Agreement, as discussed below. The transatlantic gap in this respect is an important part of the explanation.

Eppur si muove ... Notwithstanding the notoriously weak global governance architecture on energy and climate, international organizations are gradually making their voices heard. The International Monetary Fund has proposed an international carbon border price floor amongst major emitters achieved through a mix of carbon taxes, pricing and regulatory measures that entail a shadow carbon price (Parry et al. 2021). While not espousing a floor, the OECD has converged on the idea that a global carbon pricing mechanism should allow both explicit carbon taxes and cap-and-trade systems, as well as implicit measures, such as regulations banning coal-fired power plants (Fleming and Giles 2021). The 2021 G20 summit embraced, amongst several options, the idea of carbon pricing.

Furthermore, COP26 in Glasgow finally agreed on an operationalization of Article 6 of the Paris Agreement. By establishing the notion of units representing carbon emission reductions, countries would be able to trade emission credits, leading to the development of carbon markets and carbon offset mechanisms. Emitters could then buy and sell carbon credits across borders leading to the development of international carbon markets, which would spur carbon pricing at the global level. By internationally trading mitigation outcomes, in turn achieved through different mitigation policies, including cap and trade, carbon taxes and regulation, such mitigation measures would become increasingly linked. Ultimately it is the overall reduction of emissions that matters, rather than the specific policy routes to get there. Provided that each country's steps are correctly measured, with no double counting, any mitigation measure should be accepted. As different forms of explicit or implicit carbon pricing are enacted and become more widespread, countries would be encouraged to cooperate more systematically, driving up the cost of emissions. Furthermore, developed countries would be more induced to finance climate-friendly projects in emerging and developing ones. In fact, 5 per cent of the proceeds from traded offsets will be channelled to adaptation, with 2 per cent going to mitigation. While these percentages have been criticized as insufficient, they nonetheless represent a boost to climate finance, which remains well below the agreed threshold, as discussed below (Evans et al. 2021).

Achieving transatlantic common cause will likely require stepped-up US ambition as well as European flexibility. Different socio-economic conditions and political cultures entail different policy solutions, with different degrees of feasibility when it comes to carbon taxes, cap-and-trade or regulatory measures. As suggested by the IMF's proposal, so long as these measures are effective in decarbonizing, there should be mutual recognition and therefore the possibility to move forward together. Again, the transatlantic agreement on steel and aluminium indicates the way, whereby the EU and US agreed both to remove tariffs on one another and to cooperate on sustainable steel technologies. They also agreed to penalize carbon-intensive steel production elsewhere (read China) (Sandbu 2021). The logic underpinning this sector-specific climate club could be replicated in other carbon-intensive industries. Bearing in mind that the Commission's CBAM proposal is not for a blanket carbon tariff but rather revolves around specific sectors – cement, iron and steel, aluminium, fertilizers and electricity – sectoral transatlantic agreements focused on carbon-intensive industries could form the bedrock of a transatlantic climate club.

Third, transatlantic cooperation is critical to mobilize sustainable private investment. The role of the private sector in reaching net zero is essential. A few figures drive home the point. Whereas governments and institutions scrambled in Glasgow to cough up $100 billion per year on climate finance, promised in Paris in 2015 and yet to be delivered, the IEA (2021c) estimated that $4 trillion worth of investments by 2030 are needed to put the world on track for net zero. In renewables alone, IRENA (2019b) claims that approximately $27 trillion will be necessary to remain within 1.5°C of global warming. In the developed world, much of the investment is starting to take place, as the private sector is mobilized to capture the economic opportunities stemming from decarbonization, whilst being increasingly dissuaded to invest in carbon-intensive sectors. This is not happening anywhere near as much in developing and emerging countries, however. Around 70 per cent of the missing investments are precisely in these world regions, where the IEA (2021d) calculates that annual clean energy investments should increase sevenfold to around $1 trillion by 2030. Between the billions being discussed at the successive Conferences of the Parties and the trillions needed to reach net zero, particularly in emerging and developing countries, lies the private sector alongside international organizations and multilateral development banks. Without the right policy incentives, this will not happen.

This is where the EU's work on climate disclosure and especially on the green taxonomy on sustainable investments steps in. The EU revels in standard setting, and has a track record across different fields, from trade to digital privacy, health and safety, the environment and human rights. It has been described as a market, a normative and a regulatory power precisely because of this (Manners 2002; Damro 2012). Its ability to diffuse norms, rules and standards is what is commonly referred to as the 'Brussels effect' (Bradford 2020).

The ambition is for the Union to become a standard setter on social, sustainable and green investments. The prospect is within reach. The EU is already a global leader on green investments, with the EIB (2021b) reporting that, compared to the US, European companies are far more inclined to invest in climate mitigation and adaptation, with 50 per cent more patents in green technologies being registered in Europe than in the US, and four times more than in China. Spurring such investments is access to green capital. This segment, while still representing around 5 per cent of the total bond market, is growing at an astounding pace. Having jumped by 74 per cent in one year, it reached €270 billion in 2019 and €338 billion in 2020 (Leonard et al. 2021). In 2021 it topped €500 billion, with Germany and France accounting for approximately half that figure, and the Netherlands, Sweden and Spain following suit (Jones 2021). The EU's sustainable and green bond market is not only the largest worldwide, representing around half the global share, but also the most rapidly expanding one.

To boost this growth further, the EU developed a green taxonomy. To be taxonomy aligned, a project must quantify its mitigation or adaptation impact as well as demonstrate that it is doing no harm. Mitigating emissions with one hand while harming the environment or societies with the other would do no good. By establishing clear indicators, metrics and evaluation mechanisms determining green alignment, the EU green taxonomy would insert order in the flurry of market-based and civil society activities that have mushroomed in recent years. An authoritative institutional voice determining what counts as a green investment and what does not is what investors and companies would welcome, propelling this nascent market to new heights. Furthermore, as the EIB increasingly acts as the EU's implementing arm of green finance, its use of the taxonomy would contribute to spreading the taxonomy's norms beyond the EU's borders. The EIB's work with the International Platform for Sustainable Finance and its structured relations with peer multilateral development banks, as well as other public finance institutions and the private finance sector, are crucial in this respect (Erzini Vernoit et al. 2020). The taxonomy could also serve a broader public service by supporting the global development of carbon metrics and standards. This would assist countries in their efforts to set, report and revise nationally determined contributions in the context of the COP.

The process of developing an EU taxonomy was launched in 2020. It was not easy. Not only were there deep divisions between Member States, but even European institutions did not see eye to eye. The EIB, a key financial arm for many of these investments, has applied even stricter standards than those discussed by the Commission and the Council. On one level, this spurred a higher level of European ambition. On another, it added to the cacophony of messages coming from Europe.

In the EU taxonomy, agreement was relatively simple on some areas. Defining as green private investments in renewable sources, storage and batteries is a no-brainer. Yet there were other major sticking points. After much wrangling, areas like bioenergy and forestry were eventually included in the taxonomy. Strongly pushed by Finland and Sweden, the Commission eventually agreed. This triggered the ire of environmentalist groups that have claimed that such inclusion transformed the Green Taxonomy Delegated Act into a greenwashing tool (WWF Europe et al. 2021).

Converging on nuclear energy and gas was harder. On nuclear energy, divisions are stark. Some Member States, notably France, Romania, Czech Republic, Finland, Slovakia, Croatia, Slovenia, Bulgaria, Poland and Hungary, are adamant on defining nuclear energy as clean and a necessary piece of the decarbonization puzzle. Other countries, like Germany, Austria, Denmark, Luxembourg and Spain, consider it a political taboo and insist that a green investment should be sustainable throughout its life cycle, putting the spotlight on nuclear waste.

Gas was equally divisive. In the 2010s, the narrative of natural gas as a bridge fuel was often heard in European institutional, corporate as well as civil society gatherings and debates, much like in other world regions. By the end of that decade, it lost considerable steam in the EU, while remaining prominent elsewhere. However, whereas natural gas 'as is' is unlikely to return in vogue in Europe, the question is whether gas coupled with CCS should not be viewed as a key part of the picture. Countries like Germany, Italy and the Netherlands have high stakes in natural gas, but here too the resistance from environmentalist groups as well as Member States like Spain, Ireland, Denmark, Luxembourg and Austria has been adamant, claiming that investments in these sectors would divert funds away from renewables. While never explicitly recognized as the straw that broke the camel's back, the 2021–2 energy price spike pushed the European debate towards the recognition of these sectors in the taxonomy. With price volatility expected to become a structural feature of the transition, the European debate ultimately converged on the recognition that in the definition of its green taxonomy, rather than opting for clear-cut green or brown options, there should be an acknowledgement of different shades of green. In fact, rather than debating whether or not to include nuclear and gas in the green finance taxonomy, the question became how to do so.[27] While not 'green', the latter two categories would include economic activities deemed necessary to reach net zero emissions by 2050.

As the EU moves forward on the taxonomy with all the obstacles along the way, it is important that it finds common cause with the US, rather than duplicating or worse still working at cross purposes with it. Given the increasing pressure that companies, institutional investors and development banks on both sides of the Atlantic are under to raise the share of sustainable investments in their portfolios, working transatlantically to develop aligned if not harmonized criteria would serve both Europe and the US well. Alignment on a common transatlantic taxonomy has been discussed repeatedly between EU and US leaders. However, positions have not fully converged. Beyond the structural limits impeding coordination, let alone integration between different varieties of capitalism across the Atlantic (Hall and Soskice 2001), the US has indicated its reluctance to buy into the EU's green taxonomy, pointing to its general discomfort with excessive regulation. While making progress on climate risk disclosure, the Biden administration has pushed back against the notion of qualifying and quantifying what can be defined as a green investment (Mathiesen and Coleman 2021).

However, the US did not stop at going its separate way. It was also adamant on the exclusion of gas from the EU taxonomy. At first glance, this seems surprising. With the US being a major gas exporter, it seemed paradoxical for Washington to push for the exclusion of natural gas from the EU taxonomy, striking itself out. Yet on closer inspection, the US's interests emerged more clearly. Whereas the share of the US LNG in the EU is very limited, standing at 5 per cent of total gas imports, gas imports from Russia are significant, hovering around 40 per cent (Eurostat 2021). The agreement to increase US LNG supplies to Europe against the backdrop of the war in Ukraine did not fundamentally change the equation. The US push to exclude gas from the EU taxonomy is, in fact, predominantly geopolitical in nature. The US pressed for years for the reduction of European imports of Russian gas, in the belief that doing so is the prerequisite for strengthening European energy security. However, as discussed in chapters 1 and 3, the debate about energy security faded from the European debate between 2014 and 2021–2. Conflict with Russia in those years was generally discussed in relation to military deterrence, hybrid warfare, disinformation or cyber, rather than energy. Energy security, notably in relation to Russia, was overtaken by the green agenda in the EU. This influenced not only the European debate but also the transatlantic one. The unchanged US drive to push Russian gas out of the European energy equation, including the exclusion of gas from the EU taxonomy, became increasingly coloured in green.

In hindsight there was merit in the US view, particularly when it comes to the energy (in)security of some specific EU Member States vis-à-vis Russia as well as the overall vulnerability of the EU, especially in times of high energy prices. The war in Ukraine reawakened Europeans to this harsh reality. However, this does not imply that Europeans should take a blunt black-and-white approach to gas in general, including in the EU taxonomy. Doing so simply jars with reality, as amply demonstrated in 2022 when the EU and the US scrambled to increase LNG imports to Europe in anticipation of Russian gas supply interruptions. Gas, particularly as and when it is decarbonized, is set to represent an important piece of the EU's energy transition puzzle. It is difficult to imagine how the EU could do without it in the years ahead. Much like the EU would be well advised to be flexible in its drive to encourage different forms of carbon pricing worldwide, the US should be likewise on gas and its links to sustainable finance, working with the EU on a nuanced approach, which ideally would be transatlantic, foreseeing different shades of green. The EU's standard-setting power on sustainable investments would be infinitely greater if exercised in tandem with the US.

Climate finance, the Global South and European responsibility

Climate change is profoundly unjust. First, it is simply a fact that developed countries in North America and Europe are responsible for the vast majority of greenhouse gas emissions since the industrial revolution, with China – by far the largest emitter today – quickly catching up (see figure 5). The responsibility that countries bear for past and present emissions – whose effects will linger for decades given the long-term nature of the carbon cycle – is dramatically unequal.

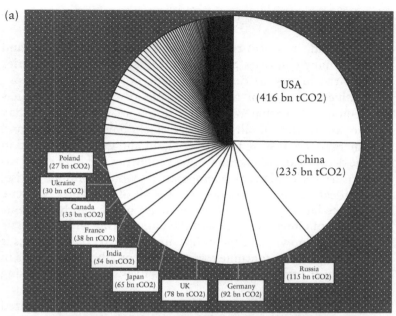

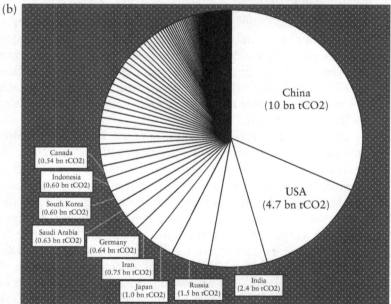

Figure 5: Historical and annual CO_2 emissions: (a) Total historical CO_2 emissions per country, 1750–2020; (b) Annual CO_2 emissions per country (2020)

Source: OurWorldInData.

Second, it is also true, unfair as it is, that emerging and developing countries cannot behave like developed countries had the luxury to do, basing their growth on the unconstrained burning of fossil fuels. Today climate science has demonstrated what we did not know a century ago. Today we know that if all countries, including developing and emerging ones, were to burn coal, oil and gas unrestrained in the decades ahead, like developed countries did in the last two centuries, global average temperatures would continue to rise, with catastrophic consequences for all of humanity. Climate science has now clarified incontrovertibly the anthropogenic nature of climate change and the damage this will cause. The facts are known, and the public and political consciousness is growing. It will continue to rise exponentially as the effects of climate change will affect everyone's lives. There is no turning back the clock.

Third, as amply discussed above, power in a decarbonized world is likely to lie predominantly in the hands of the most powerful countries. As value moves down the supply chain from commodities to technologies, and as the major powers push the accelerator on the energy transition, the gap between the haves and the have-nots could widen (Goldthau 2017). In fact, the risk is that as we address climate change, we may witness new forms of extractionism and colonialism in the Global South.

Finally, the effects of climate change itself are terribly unequal. All countries and peoples will be affected by increases in average temperatures, rising sea levels and a greater frequency and intensity of extreme weather events. Studies highlight that up to 3 billion people out of the projected 9 billion in 2050 could end up living in areas whose average temperatures will fall beyond the range in which humanity has thrived for the last 6,000 years (Xu et al. 2020). With each additional degree of warming, 1 billion people could end up moving out of what is known as the 'climate niche', that is, those areas of the planet that have seen civilizations emerge, flourish and consolidate over the millennia. Most of these people live in the Global South, in Asia, Latin America and, especially, in Africa. Added to this,

Africa is and will experience a demographic explosion in the years ahead. This makes the climate crisis and access to energy, which remains woefully insufficient today, mutually reinforcing wicked problems. African countries face the challenge of providing energy to an exploding population – of which 600 million still lack access to electricity today. To do so, the most readily available solutions still involve the burning of fossil fuels. At the same time, they bear the brunt of the effects of climate change, that is generated precisely by the burning of those hydrocarbons, largely elsewhere. Today Africa is responsible for under 4 per cent of global emissions. In other words, not only is the Global South not to blame for climate change, both historically and today, but it is expected to pay the heaviest price of its devastating effects in future.

It thus comes as no surprise not only that the principle of 'common but differentiated responsibilities' has been a staple of the international climate consensus since the 1997 Kyoto Protocol, but also that climate justice, including 'loss and damage', has been one of the thorniest issues in international climate negotiations and an unbridged cleavage between the Global North and South. Initiated in 1991 by the Alliance of Small Island States calling for compensation for sea level rise, the notion of loss and damage first appeared in the 2007 Bali Declaration and was then outlined in the 2013 Warsaw International Mechanism on Loss and Damage. Eventually, it made its way into Article 8 of the 2015 Paris Agreement, and thus became enshrined as a thematic pillar of the United Nations Framework Convention on Climate Change (UNFCCC).

However, the Conference of the Parties has struggled to translate this principle into practice. The Paris Agreement, while recognizing the notion, was quick to specify that loss and damage 'does not involve or provide a basis for any liability or compensation' (Decision 52 of the 2015 Paris Agreement). Within and beyond the agreement, developed countries have adamantly resisted the operationalization of the principle, which could expose them to legal claims for liability or compensation. The sums involved are indeed staggering. Research suggests that climate-related damage in the least developed countries and small island states could be as high as 64 per cent of their GDP by 2100 under current climate policies. Africa would bear the brunt, with Sudan topping the list, with an 84 per cent drop of GDP. This would translate to hundreds of billions per year worth of compensation (Andrijevic and Ware 2021), without considering unquantifiable damage, such as the destruction of cultural heritage or the degradation of ecosystems. To date, only Scotland has accepted to earmark a largely symbolic $2.7 million fund for 'loss and damage' at COP26 in Glasgow. No other developed country has followed.

Whereas COP26 has at least agreed to begin a dialogue on loss and damage, its operationalization remains over the horizon. In other words, the developed countries resist recognizing concretely their responsibilities when it comes to compensation for damage that cannot be prevented or adapted to. What developed countries have acknowledged is their responsibility to provide climate finance to emerging and developing countries, both for mitigation as well as adaptation. They have accepted partially their political – albeit not legal – responsibility for climate injustice. The 2015 Paris Climate Agreement set out a concrete target: richer countries pledged $100 billion per year to support climate mitigation and adaptation in poorer ones. While there remained anger and acrimony in the climate justice debate, at least the world had a concrete number to work towards.

Climate finance remains a thorny issue. The years have gone by, and while climate finance has increased, the $100 billion mark is still beyond reach. The OECD estimated that by 2019 developed countries had provided $80 billion in climate finance, up from $78 billion the previous year (OECD 2021). This is encouraging, both as far as the upward trend is concerned and the absolute numbers. In the run-up to COP26 in Glasgow, several countries increased their pledges on climate finance, including the EU, the US, Germany, Canada, Japan, Italy and Germany. In addition, through the Glasgow Financial Alliance for Net Zero, over 450 companies from forty-five countries committed to mobilize $100 trillion of private capital to reach net zero emissions.

Others have expressed scepticism of the current state of play though (Timperley 2021). The OECD is essentially composed of those same developed countries expected to cough up the cash, and its reports are grounded on the data provided by them. Many of these funds are loans rather than grants, and part of the grants is carved out from existing development assistance rather than being added to it. Civil society groups like Oxfam (Carty et al. 2020) have calculated that the actual figures may be as low as one third of those given by the OECD. When countries pledged important increases – such as the US, which in the run-up to COP26 promised a doubling of its climate finance – it is not entirely clear how much of this would be new rather than repurposed or, worse still, relabelled, money. Inflated figures aside, even under the most generous estimates, the world remains far from meeting its climate finance target. COP26 had set the ambition to ramp up climate finance to meet and ideally go beyond the Paris pledge. While the Global North tries to cobble together $100 billion per year, in fact, the Global South debates radically different figures, with the South African environment minister suggesting $750 billion annually by 2030 (Timperley 2021). The global cleavage remains deep. Currently the world is set to meet the $100 billion only by 2023 at the earliest, and most probably later. Going beyond that target remains an even more distant goal.

Regarding the climate funds that are on the table, further questions revolve around how they are spent. Specifically, there has been a strong bias in favour of projects aimed at mitigation over adaptation. The OECD found that of the estimated $80 billion of climate finance in 2019, only $20 billion were spent on adaptation (OECD 2021). There are several reasons for this.

First, climate adaptation for many years was viewed as a dirty term, somehow conveying a sense of defeatism, or worse, the implicit message that money should be spent to adapt to a hotter world, rather than to prevent it. Climate activists themselves struggled to swallow the notion of adaptation, notably deep adaptation (Nicholas et al. 2020). Only recently has the awareness grown that even in the best of circumstances – i.e., meeting the 1.5°C goal – the effects of climate change will be significant and imply a very heavy cost. If, starting tomorrow, emissions were cut by 5 per cent annually, the positive effects in terms of climate would start being felt only after a couple of decades. In the meantime, average temperatures would continue increasing alongside rising sea levels and extreme weather events. As a consequence, Andrijevic and Ware (2021) estimated a 33 per cent drop of GDP in less developed and small island states by 2100 even if the 1.5°C goal is met. Current mitigation measures will not reduce, let alone eliminate, existing disruptions, but only slow down the rate at which these worsen in future. The UN has estimated that developing countries already need $70 billion per year to meet rising adaptation costs, with this figure rising to $300 billion over the next decade (UNEP 2021).

Second, public donors tend to prefer mitigation over adaptation projects because their outcomes are easier to quantify. Complex as it is, it is easier to measure the reduction of emissions generated by a specific decarbonization initiative rather than the success of a particular adaptation project. The latter often entails imagining counterfactuals such as the prevented cost of a possible extreme weather event, which hardly passes the test of scientific rigour. Given the need for political accountability for how taxpayers' money is spent, as well as the institutional logics driving development policy, with their strong emphasis on measuring impact, public funding has veered towards mitigation projects, whose effects can be more easily identified, accounted and showcased.

Third, companies and financial players prefer mitigation over adaptation in their investments given the stronger business case for such projects. Decarbonization is no longer simply associated with bans, costs and restrictions. It is also increasingly associated with economic opportunity. As European leaders tirelessly repeat, growth and greenhouse gas emissions must and can be decoupled. The private sector understands this. We are still well below the $4 trillion of annual investments in decarbonization by 2030, which the IEA (2021a) estimates are necessary to reach net zero emissions. However, the bulk of the private investments that do exist focus on mitigation, be it in terms of low emissions fuels, energy efficiency and end-use, or clean electricity, and are geographically concentrated in developed countries. There is a strong business case for many of these mitigation projects. Adaptation is a different ball game. Whereas the challenge of mitigation can be largely solved by the market, provided the right institutional, regulatory and financial incentives are put in place by public policy, adaptation represents an even greater market failure, that will require a far more significant government role (Steele n.d.). The profitability of adaptation, particularly in the Global South, simply does not match that of mitigation investments, especially in the Global North. This means that when it comes to adaptation, the lion's share will need to come from public finance, including states, international bodies such as the EU, as well as multilateral development banks. An important and largely underreported step forward made in Glasgow was precisely the unpacking of climate finance into mitigation and adaptation components, with developed countries being 'urged' to 'at least double' their adaptation funds by 2025 (Washington Post 2021).

Globally, the climate finance picture is patchy at best, with the richer and dirtier Global North falling well short of its obligations towards the more vulnerable Global South, which is both infinitely less responsible for the climate crisis and more affected by it. Major economies like the US, Canada and Australia contribute less than half their fair share, based on the size of their economies and greenhouse gas emissions (Bos et al. 2021).

Within this unedifying global picture, Europeans can at least pride themselves on being the best of a bad lot. President Biden pledged to raise annual climate finance to $11.4 billion by 2024, making the US the largest single climate finance contributor. However, even assuming this figure passes the test of Congress, which is a big if, this is less than half the funds that the EU collectively contribute, notwithstanding the roughly comparable economic and demographic weights of the two, as well as the US's greater responsibility as far as historical and current emissions are concerned. The EU, including Member States and European institutions, provided €23.39 billion in annual climate finance in 2020, and Member States like France, Germany, Austria, Sweden and Denmark already contribute more than their fair share (European Council 2021b). Furthermore, the Commission pledged to raise its climate funds by €5 billion by 2027, while Germany and Italy, amongst others, also committed to significant increases in the run-up to COP26 (European Commission 2021j). If one adds the UK, which has pledged to double its international climate finance to at least €13.8 billion per year by 2025, Europeans will come close to contributing half the global climate funds in the race to $100 billion (COP26 2021).

Furthermore, whereas the lion's share of global climate finance tilts towards mitigation, the same cannot be said for the EU, where as much as 52 per cent of all climate funds are spent helping third countries adapt to climate change (European Commission 2021e). This is not true for all Member States, with France and Germany heavily skewed towards mitigation, as opposed to Belgium, Ireland or the Netherlands that come closer to the 50 per cent mark (Bos et al. 2021). However, the fact that the EU as a whole spends over half its climate funds on adaptation is important. Beyond public grants, the EU is also playing a key role in spear-heading climate action amongst development banks, with a €1 trillion climate bank roadmap making the European Investment Bank both a powerful lending arm in the service of the European Green Deal, and the first multilateral development bank to become 'Paris-aligned' (Farand 2020). This includes a significant uptick in climate adaptation funds to 15 per cent of the Bank's finance by 2025, ending the financing for fossil fuel energy projects, and accelerating the financing of clean energy innovation, energy efficiency and renewables.

The pledges, particularly compared to other major powers, are impressive. The challenges concern implementation. Most of the Commission's climate finance for the Global South is drawn from the EU's unified external action fund, the Neighbourhood, Development and International Cooperation Instrument (NDICI). Whereas previous budgetary cycles had seen a proliferation of specific geographic and thematic external action instruments, including enlargement, neighbourhood and development cooperation, a major innovation of the 2020–7 Multi-Annual Financial Framework was the pooling of all such funds within NDICI, with an aim to increase simplification and flexibility. Within NDICI, 30 per cent of the funds are destined to climate and the energy transition. Adding up the green components anticipated under the geographic, thematic and rapid response pillars of this single instrument, this amounts to around €25 billion over seven years. Geographically, most of this money is destined for Africa, which will receive around three quarters of the total funds available (European Commission 2020a).[28] Furthermore, part of the NDICI will be channelled to a €53.5 billion European Fund for Sustainable Development, aimed at triggering private sustainable investments, thus replicating the success of previous initiatives such as the 2016 external investment plan of the European Commission. The EU estimates that over half a trillion euros' worth of private investments could be mobilized in this way.

These are significant sums, especially compared to the past. During the 2014–20 budgetary cycle, in which the EU's external financial instruments were split into different geographic and thematic components, the Union had included specific climate initiatives. In particular, after the 2015 Paris Agreement, the EU upgraded its Global Climate Change Alliance Plus to assist least developed countries and small island states. The Alliance had raised €750 million by 2020 (GCCA+ 2021). The sums discussed today are of a totally different order. The substantial earmarked EU funds for climate mitigation and adaptation raise the expectation of a new page being turned in the EU's support for climate finance, especially adaptation, to the Global South.

However, meeting such expectations will not be easy. Given that the geographic – rather than the thematic – pillar of NDICI represents the largest share of the instrument, it is not clear how to ensure that a third of these funds will be channelled to climate, biodiversity, environment and wildlife management in practice (Climate Action Network 2021). Insofar as programming is made together with beneficiary countries, the latter may press for funds being spent on other priorities instead. This includes third party governments that, while publicly calling for climate adaptation, may then push for EU funds to be spent elsewhere. This risk highlights the broader downside of the European Green Deal, and particularly the 'Fit for 55' package, which is essentially designed as an internal set of policies, with a rather underdeveloped and vague external dimension. Hence, whereas specific climate and transition funds are being established with a view to being spent internally – i.e., the Just Transition Fund and Climate Social Fund – the same does not apply externally where only a general percentage is indicated, with the specific implementing mechanisms remaining to be worked out along the way. Today many of the specific financing mechanisms, on which the Global Climate Change Alliance Plus depended, have been merged into NDICI, making it unclear whether and how this initiative will persist in the years ahead. In fact, there are no thematic external climate and transition funds foreseen.

When discussing the internal dimension of the European Green Deal, and in particular the need for EU funds to address the potentially regressive socio-economic effects of the transition within the EU, chapter 2 argued that the Just Transition Fund and the Climate Social Fund represent only the tip of the iceberg of the overall funding that will be necessary to make decarbonization socially and therefore politically sustainable within the EU. 'Greening' all other funding instruments, and thus mainstreaming climate and the energy transition across all, or most, funding mechanisms will also be necessary. In the case of the EU's external action, and especially in its external climate finance, the reverse argument can be made. The EU is right to designate a significant percentage of its unified external funding instrument to climate mitigation and adaptation. However, to ensure that 30 per cent of all funds are actually spent on these priorities, developing specific mechanisms and programmes is probably necessary. This is especially because of the role played by third states in influencing how EU development funds are spent and technical assistance is deployed. Much like the EU has established specific mitigation and adaptation initiatives internally to prevent the potentially regressive socio-economic effects of the transition and thus ensure its political acceptability, it will need to develop parallel external measures to make the transition geopolitically sustainable and avoid deepening further the climate cleavage between the Global North and South.

* * * * * * *

The twenty-first-century international system is cursed with old and new geopolitical cleavages, namely the 'old' one between the Global North and South, and the crystallizing rivalry between the US and China, or more broadly, between democracies and autocracies. In fact, whereas the Global South actually includes countries at very different stages of their economic development – such as China and most of sub-Saharan Africa – the geostrategic rivalry between the West and China (and Russia) artificially lumps all non-Western countries into the same box.

Energy has always been an important variable shaping international affairs. However, in the world in which fossil fuels dominated the scene, energy was somewhat of a sideshow. With the exception of acute crises, most prominent of which is the 1973 OPEC oil embargo in response to the US support for Israel in the Arab–Israeli war that year, energy geopolitics and the core issues dominating international relations have only occasionally overlapped.

This is far from being true now and in the years ahead, over the course of the energy transition and in a decarbonized world that it is hoped will follow. Power in energy terms will increasingly coincide with power broadly construed in the international system, given the role played by the economy, industry, technology and governance in the production, management and consumption of decarbonized energy. This has two implications. First, climate, while recognized as the quintessential global public good that requires multilateral cooperation over and above geopolitical divides, may well end up being caught up in them. This is particularly true of relations between the US, China as well as the EU. Second, climate and energy risk widening the gap between the haves and the have-nots, exacerbating the divide between the Global North and South. The frustrations and grievances about climate injustice that have long hampered progress at the successive Conferences of the Parties could grow exponentially as the effects of such injustice are increasingly felt, while being instrumentalized by players like China and Russia for geostrategic reasons that are far removed from climate action.

Within this complex picture, the EU has a powerful story to tell as an established and ever more ambitious climate leader, but one that could easily be drowned by the divisions, disagreements and conflicts about and around the energy transition worldwide. To avoid falling into this trap, the EU must begin by recognizing – not just in words but in policy practice too – the global implications and ramifications of its role in the global transition, developing far more significant and specific external policies in response. These will involve diplomacy and development, industrial policy and international finance; they will be bilateral as much as plurilateral and multilateral. In short, they will involve an all-of-government approach by Member States and the EU as a whole.

Conclusion

The green and global sides of the European coin

This book has made a straightforward case. Europe has set the goal of net zero greenhouse emissions as its new narrative and mission. After decades in which it had lost its way, torn by political divisions and policy paralysis, the EU has elevated the European Green Deal into its normative vision, strategic imperative, economic growth strategy and route to a political Union. First, a green Europe represents a normative vision: the EU has identified the climate crisis as the major challenge of our time and the energy transition as the necessary path to address it. By pursuing decarbonization, the EU recognizes the existential nature of the crisis faced by people and planet alike. Second, a green Europe represents a strategic imperative to achieve energy security: blended into the short-term need to diversify fossil relationships, an accelerated decarbonization trajectory is the only way to ensure the EU's long-term autonomy. Third, a green Europe represents a growth strategy: it does not imply only restrictions, bans and rules, words with which the European project has too often been associated. While these are necessary, equally if not more important is unleashing the creative and productive potential of the energy transition: CO_2 emissions and GDP can be decoupled. Achieving net zero greenhouse emissions will mean fostering innovation, promoting industry and creating jobs. A green Europe thus can and must be a growth strategy. Finally, a green Europe is a political project: addressing climate change is a priority that Europeans share. Unlike the poisonous divides that ripped the EU apart over austerity and migration, Member States, notwithstanding their differences, agree to tackle the climate crisis and do so together. The European public is even more adamant about this goal, seeing in climate action the first public policy priority that leaders should address. Hence, alongside being a normative vision and an economic growth strategy, a green Europe can also reconcile Europeans both to one another and to the European project as a whole.

All this is possible if a green Europe succeeds and therefore materializes in practice. This is not automatic; it cannot be taken for granted. This energy transition will happen, of this we can be sure. It is not the first, nor will it be the last energy transition that humanity has experienced. However, the difference between all previous transitions and the one we are undergoing now is 'time'. In the past, transitions were essentially driven by technological innovations and market forces, which slowly and erratically led societies and economies to adapt, eventually overcoming political, economic, social and cultural roadblocks and resistances. This time, alongside technology and markets, what is driving the transition is the growing political and societal awareness of the anthropogenic nature of climate change and its catastrophic effects for all. This means that politics and policy must accelerate the pace of change and ensure that, while taking differentiated paths and reaching the finishing line at different points in time, there is a single net zero destination for humanity as a whole.

The political and policy-driven nature of this energy transition has implications for all countries and regions. It affects the United States as it seeks to bridge the acrimonious climate policy divide between Democrats and Republicans. It concerns China, which at once strives to consolidate its primacy in renewable technologies while fuelling an energy-hungry and still coal-dependent economy. It is essential for India, whose growth has picked up yet struggles with air pollution and finding ways to be powered by non-fossil sources. It is crucial for Africa, which grapples with the twin challenges of energy access and decarbonization, having the vision but lacking the means to power its development, amidst a ballooning population, whilst bypassing fossil fuels.

However, the implications are perhaps greatest for Europe. This is because Europe as a whole – including the EU but also non-EU members such as the UK and Norway – are by far the most advanced along their energy transition journey. Unlike other players that pledge to reach net zero and at most enact a few timid steps in practice, Europeans are mobilizing funds, passing regulations and tightening pricing and taxation measures aimed at creating the institutional, financial and legal frameworks and incentives to make decarbonization happen. While not the only possible plan nor perhaps even the ideal one, the European Green Deal is the only available plan that charts the way from today to a net zero continent by 2050.

This primacy has pros and cons. On the upside, the EU is recognized worldwide as a global climate leader, ratcheting up the level of climate ambition in international climate negotiations over the years. Its credibility is affirmed not just by words but by action too. While more could be done, Europeans put to shame other developed economies in their commitments to climate finance, and in particular climate adaptation. As a trailblazer on issues like carbon pricing and the green taxonomy on sustainable investments, the EU is a model for others to study and eventually emulate. While no other country[29] has a cap-and-trade or carbon tax system displaying the same level of ambition and sophistication as the European ETS, dozens of countries have moved or are planning to move in the same direction as the one charted by the EU since 2005. The EU green taxonomy is still embryonic, but here too, as sustainable investments grow, companies and investors will increasingly strive for a clear and established set of rules. To the extent that the EU is the first major institutional mover in this field, it is likely to represent a standard for others to follow.

However, being a first mover also comes with risks and costs, especially if others do not follow. The energy transition, akin to a revolution, can and will represent a net gain for Europe and for the world. This is true when viewed from different angles, be it environmental, social, security or economic ones. However, like all revolutions, this transition will feature winners and losers, both within the EU and beyond European borders. Policy is thus tasked with the complex goal of accelerating the transition, magnifying its gains whilst compensating its losers both internally and internationally. Doing so will not be easy. The task is complex but feasible if the rest of the world moves in the same direction, albeit along different paths and at different paces. If the world moves as one, the EU, leading the way to net zero, can weather the energy transition storm both at home and abroad.

By contrast, if Europe moves to decarbonize itself but others do not follow, or if the transition gap between the EU and the rest of the world widens too much, the resistances the Union will face both internally and internationally may be too great to endure. Internally, the EU risks a backlash from companies and lower-income households faced with intolerably high energy costs. If the EU's energy transition drive is unmatched by others, the social resistance from carbon-intensive industries, small and medium-sized companies, lower-income households and coal workers could translate into political paralysis and setback. Internationally, if the EU's leadership is not followed by the rest of the world, rather than a decoupling of emissions and growth, the risk is that of a decoupling of the green and brown economies. Europe would find itself in splendid isolation, unable to navigate the triangle between energy security, decarbonization and affordability. If left alone, Europeans would have to grapple with the geopolitical disruptions that its energy transition might unleash in its surrounding regions, while also bearing the consequences of the climate crisis itself, exacerbated by insufficient climate action by others.

Given the exquisitely political and policy-driven nature of this energy transition, addressing its social, economic and therefore political and geopolitical consequences is essential. An ill-managed European transition, which does not do so and which fails to draw the rest of the world with it, could end up being catastrophic. It would be catastrophic because of the social and economic disruptions it would cause both internally and internationally. These would reverberate politically, putting at risk liberal democracy, as well as geopolitically, by unleashing security, economic, social and political disruptions that would boomerang back to Europe. Whereas a green Europe signals a revival of the European project, a failed, stalled or reversed European energy transition could seal its demise. An ill-managed transition would be disastrous also because it would turn back the clock on Europe's net zero goal and on the rest of the world's by representing an example for others not to follow. In short, whereas a successful European energy transition holds the promise to revive Europe politically and economically, while contributing to saving the planet, a failed transition means exactly the opposite. The stakes could not be higher.

Reconciling institutionally a green and global Europe

The EU has embarked on its decarbonization journey with courage and conviction. It is a long and complex journey that will last at least three decades, likely more. There will be many hurdles to overcome, trade-offs to face and compromises to make along the way.

Rising to the challenge, the EU is equipping itself internally. The European Green Deal, beyond representing a concrete plan to reach net zero greenhouse gas emissions, also entails an all-of-government ecological approach. It features the prime goal of decarbonization, with the industrial, financial, regulatory and fiscal measures to achieve it, but covers also broader environmental matters, from biodiversity, forestry, pollution, food, land use and agriculture. Internally, the EU is adapting its institutional set-up to reflect this approach. Whereas the Commission Directorates General (DGs) CLIMA, Environment and Energy existed before, today climate and energy permeate many other DGs of the European Commission. The first Vice President of the

European Commission is explicitly tasked with the mandate
to see the European Green Deal through. Likewise, whereas
climate and energy matters have long been on the agenda
of the Council of Ministers of the EU, notably through its
environment, and its transport, telecommunications and
energy configurations, today climate and energy matters spill
into many other Council formats, and often make it onto the
agenda of the European Council as well. Between 2014 and
2021, energy featured on average twice and climate three
times per year on the agenda of the monthly meetings of the
European Council, with peaks in 2014 and 2021 against the
backdrop of energy price slumps and spikes respectively. The
same is true of the European Parliament, with the Committee
on the Environment, Public Health and Food Safety, and that
on Industry, Research and Energy focusing on climate and
energy matters, and these increasingly permeating the work
also of other committees, including budget, economic and
monetary affairs, employment and social affairs, transport,
agriculture and development, amongst others.

More could be done at both EU and Member State level to
break down institutional silos and put climate and the energy
transition at the centre. First, maintaining a separation
between institutions dealing with climate and energy is
artificial at best, counterproductive at worst. Given that
three quarters of climate action passes through the energy
transition, it is hard to understand how concrete progress can
be made without greater institutional synergies, if not mergers,
between the two. Institutional separation between energy and
climate leads to very different, and at times contrasting and
often non-communicating, institutional cultures, networks
and policy approaches. For instance, when the climate and
the energy policy communities work on the external dimensions of their respective policy fields, they end up focusing
on very different issues. Climate officials concentrate almost
exclusively on multilateral climate negotiations in the context
of the UNFCCC, whereas their energy peers work on energy
security and the functioning of international energy markets.
Both are important and actually intertwined, yet in practice
they proceed along parallel planes. The strategic cost of such

separation was made abundantly clear in light of the war in Ukraine, in which institutions, politics and public debate more broadly scrambled to reconcile the narrative, policy and practice of energy security with those on the energy transition. In some Member States, such mergers have happened and the results are encouraging. In Italy, for instance, Mario Draghi's government for the first time merged climate, energy and economic development in a single ministry for the ecological transition, which in turn became one of the main 'power ministries' of his executive. Other Member States, as well as the European Commission, that still artificially separate climate and energy, could move in the same direction.

Second, whereas climate and the energy transition increasingly feature in the work of different institutions, they are yet to become priorities for those institutions that are not expressly tasked with these matters. It is critical for this to happen, especially when it comes to economic, finance and trade institutions, through which climate mitigation and adaptation largely pass. Given that climate and environment institutions are generally not where 'power' truly lies, it is crucial that climate and the energy transition start occupying centre stage of those institutions – like finance and the economy – where the most consequential decisions are normally taken.

Third, central institutions like presidencies and chancelleries should play a coordinating role to ensure the effective mainstreaming of climate and the energy transition across government. In this respect, the reorganization of the European Commission with an Executive Vice President overseeing the EU's climate action is an example for Member States to consider and follow.

Internally, the EU is starting to walk the walk of a green Europe. It is still at the beginning of the journey, but important steps have been taken both in Brussels and in several European capitals. Internationally, Europe as a whole is simply not where it should be, however. The European Green Deal is translating into an all-of-government approach but this is essentially confined to institutions and policies covering internal policy matters within the EU. We remain a long way away from extending this all-of-government approach to the EU's external action and therefore realizing in practice what a green and global Europe implies.

This is not to deny that some steps forward have been made. The 2003 European Security Strategy did not once mention climate change or the energy transition (European Council 2003). In fact, when energy was discussed in the context of foreign and security policy in the early 2000s, it was mainly in terms of the need to access cheap and reliable hydrocarbons to ensure Europe's energy security. As a major consumer but not producer of oil and gas, the European energy–foreign policy nexus revolved around ensuring greater access to fossil fuels. This approach reflected the broader academic and policy debate in those years, whereby the geopolitics of energy was basically understood as the geopolitics of fossil fuels, not of climate change nor of the energy transition.

Whereas the awareness of climate change back then was certainly not what it is today, what is striking, in retrospect, is that this was the case even though international climate negotiations had been ongoing for several years and the EU had emerged as a key player in them. The EU's first security strategy, published six years after the Kyoto Protocol and two years after the US's withdrawal from it, de facto handing over the mantle of climate leadership to Europeans, did not feature climate change or the transition at all. Climate diplomacy played out in its separate bubble, and climate change and the energy transition were entirely absent both conceptually and practically from European foreign and security policy thinking.

In the first decade of the twenty-first century, the first timid steps to integrate climate change into European foreign policy were made (Oberthür and Dupont 2021). The EU moved to speak with one voice, with the rotating EU presidency representing the Union in international climate negotiations, the European Commission providing the necessary support and expertise to substantiate the EU's climate objectives and strategies, and the green diplomacy network established to coordinate the climate initiatives of Member States and EU institutions. The failed Copenhagen Conference in 2009 spurred the EU further into external action. Beyond engaging in the UNFCCC framework, Europeans began reaching out more systematically to third countries, regional organizations and transnational climate initiatives such as the Covenant of Mayors. After the Copenhagen debacle, the EU understood that its climate ambitions could carry the day only if embedded in broader climate coalitions and consensuses, pursued through institutional and informal channels, within transnational, bilateral, mini-lateral and multilateral settings.

The Treaty of Lisbon and the establishment of hybrid institutions such as the triple-hatted High Representative, Vice President of the Commission and Chair of the Foreign Affairs Council, as well as the European External Action Service staffed with Commission officials and Member State diplomats, changed the dynamic further. Unlike the 2003 European Security Strategy, the 2016 EU Global Strategy, which this author penned on behalf of the High Representative and Vice President (HRVP), mentioned climate and energy twenty-one and thirty-six times respectively. Climate and energy had made it on to the foreign policy map by the mid-2010s. Today there is far more coordination and coherence both horizontally between institutions in Brussels and vertically with Member States on climate and energy matters. Until 2018 the European External Action Service (EEAS) did not deal with climate and energy at all, whereas today it includes an ambassador at large for climate diplomacy and a division responsible for the green transition. The Foreign Affairs Council periodically turns its attention to climate and energy matters. Some such occasions, such as in January 2021, were instrumental in building European consensus on key issues such as the need to put a halt to unabated coal production and consumption, enabling the EU to push this goal, albeit with partial results, in Glasgow later that year. The European Commission, and in particular its first Vice President Frans Timmermans, represents the point person in international climate talks and the main counterpart to the US or the Chinese climate envoys, amongst others. In short, important steps forward have been taken in recent years on the way to a green and global Europe.

This said, there is a very long way to go. Within the EEAS, there are only five officials plus an ambassador at the headquarters working on the energy transition. For a service that should represent the global face of the EU's new green raison d'être, this hardly communicates a sense of priority. Some Member States, notably Germany, France and the Nordic countries, are well ahead of the game in this respect. In Denmark – with a population of under six million – the ministry of foreign affairs has twelve people working on climate and energy, more than double the number in the EEAS. In Germany, alongside a climate envoy, there are approximately thirty officials dealing with climate and the transition in the Auswärtiges Amt, many of whom transferred from the environment ministry. Given that 92 per cent of emissions are produced outside the EU, the contrast between the six officials in the EEAS and the thousands in the Commission working on the internal dimension of climate and energy is stark.

Likewise, the Foreign Affairs Council now turns its attention to energy and climate matters, but does so only periodically. In the sixteen meetings held in 2019, energy featured twice and climate four times. In 2021, numbers increased, with energy featuring five times and climate seven times out of the fourteen meetings held.[30] The trend is encouraging. However, walking the walk of a green and global Europe would imply much more. The regional and global implications of climate change and the energy transition ought to become regular topics of in-depth discussion and ensuing action by European foreign ministers.

Equally important is the joining up of institutions focusing on the internal and international aspects of climate and energy. Joint meetings between different Council configurations are never easy with policy silos and siloed mentalities being hard to bend. When migration was viewed as the be-all-and-end-all of EU policy-making in the mid-2010s, various attempts were made at joining up the internal and external dimensions of EU migration policy. The ensuing joint meetings between the internal and foreign affairs bodies of the Council had mixed results at best (Tocci 2019b). Greater success was achieved in joining up the work of the Commission and the EEAS in those years, for instance on multidimensional priorities such as resilience building (Tocci 2019c), as well as creating smoother cooperation between the European Parliament and the HRVP.

However, the institutional tide has turned since then, and not for the better. Institutional turf wars, while never having disappeared, are on the rise again in the Brussels bubble. This is true of relations between the Commission and the European Council, and it is evident in the increasingly subdued role of the EEAS, which was meant to represent a hybrid institution between the two. One could go as far as arguing that the EU, having overcome successive crises in which different institutions took the upper hand at different points in time – the European Council and the Council of Ministers in the eurozone and migration crises, and the Commission during the Covid-19 pandemic and the Ukraine war – is adjusting institutionally to a post-Lisbon Treaty reality. The experiment of hybrid institutions, despite their promise and potential, is on the wane. This does not imply an end to such institutions. So long as the EU does not engage in a revision of its Treaty – unlikely to happen anytime soon – the current institutional set-up will remain in place. However, this does not prevent a changing power dynamic between institutions, with some gaining the upper hand and others being sidelined or hollowed out along the way.

The institutional context may not be promising but the political imperative is stark. The months and years ahead will need to see much greater attention devoted to climate and the energy transition by foreign policy institutions like the EEAS, the Foreign Affairs Council and European ministries of foreign affairs, more structured horizontal coordination between these bodies and their internal policy counterparts, as well as more systematic vertical coordination between national and European institutions. Greater institutional coordination can positively influence those bodies that lag behind in terms of staff and resources devoted to energy and climate. Working more systematically with peers that have already made the strategic choice and ensuing financial and organizational adjustments to prioritize climate and energy could speed up similar changes in their own institutional homes. At the same time, systematic coordination could magnify the global role and impact of a green Europe. While it is true the EEAS only features a handful of people working on climate and energy, when Member States are added to the count, the number of European officials working on international climate and energy matters rises to approximately seventy. Harnessing such potential through structured and systematic coordination is crucial.

Finally, the foreign policy community at national and European levels will need to work in greater sync and synergy with civil society, the media and the private sector on the transition. Given that the transition implies fundamental changes in how we buy, work and live, it is crucial to keep civil society and the media informed and engaged in order to bring the public along what promises to be a tumultuous journey in the months and years ahead. As the transition starts playing out in practice, with winners and losers coming to the fore, their role will be essential in fostering European resilience, cohesion and determination to follow through with its climate and transition ambitions. The private sector, in particular the energy one, but also European industry and finance more broadly, represent the link between European ambition and actual results. The energy transition implies a fundamental transformation of our industries and economic development models. Without the private sector, that transformation will simply not happen, undermining the EU's internal political agenda and its international credibility. Official institutions can accelerate and steer the energy transition effectively only by working hand in hand with the private sector and civil society.

From climate security to a green and global Europe

A green and global Europe requires an overhaul of policy-making. In recent years, several authors have pointed to the need to integrate energy and climate matters organically into European foreign policy. Approaching the question from a climate perspective, they point to the existential nature of the climate crisis, which implies that all actors, including foreign and security policy ones, should focus their minds on the question (Proedrou 2020). Rather than addressing only or mainly 'traditional' international politics and security, foreign policy should concentrate on the planetary threat posed by climate change, given the security threats and challenges unleashed by its consequences. Indeed, the security disruptions caused, triggered or magnified by climate change are already here, with the role played by the food price spike in the Arab uprisings and the Syrian civil war being the most cited examples (Bergamaschi et al. 2016). These represent only a small foretaste of the disruptions in store. The future will increasingly see climate-created or magnified conflicts within and between states, food insecurity, global trade disruptions, humanitarian crises and migratory waves (Youngs 2015). Foreign and security policy should thus adapt, integrating and mainstreaming 'climate security' into its thinking and practice (Bordoff 2020; Youngs 2020).

Viewed from this angle, the foreign and security policy community has started responding. Back in 2008, High Representative Javier Solana put the spotlight for the first time on the climate–security nexus in a policy paper to the European Council (2008). It took several years for foreign ministers and ministries to respond, but today the idea that climate change represents a security threat multiplier is well understood in these institutions. The defence policy community woke up later, but here too there is now some movement to report. In 2019 European defence ministers discussed for the first time the climate–defence nexus and have since then started working on how they should restructure their work to better anticipate crises caused or magnified by climate change, re-tailor their operations, as well as adapt their capabilities to warmer climates.[31] This is happening in the US too, within the Department of State and the Department of Defense, as well as in security and defence players like NATO. The same is true of the foreign policy community in general, from academia to civil society and the media, both in Europe and in the wider world. Major events such as the Munich Security Conference regularly and increasingly feature discussions on climate change.

This book makes a far more fundamental claim. Yes, climate change has critical security and even defence implications. These must be factored into foreign and defence policy-making both in Europe and elsewhere. But climate change and the energy transition go much deeper and wider. They affect society, industry, finance, technology, security and the economy, lying at the core of the future of our liberal democracies at a time when they need to demonstrate their enduring worth. Climate and the energy transition will determine the prosperity and security of present and future generations, and they will fundamentally shape Europe's role in an ever more contested and connected world. The existential nature of the climate challenge alongside the magnitude and complexity of the energy transition require an all-of-government approach spanning internal and external policies and institutions, as well as civil society, academia, the media and the private sector.

Europe is far from featuring seamless cooperation between policy sectors and institutions, with institutional turf wars and silos still being staples of European political and policy diet. European policy compartmentalization does not stop here, with revolving doors between institutions, academia, civil society and the private sector, rather than being seen as goods to be cherished, are more often considered ills to avoid. Entrenched interests, institutional inertia and political cultures all militate against a European all-of-government approach to the transition already underway.

Yet Europe has identified the green agenda as its new guiding light. As this book has argued, it has done so for good reason. A green Europe holds the key, strategically, socially, economically and politically, to Europe's revival. The die on a green Europe is cast; from this, there is no turning back. Now action must follow, internationally as much as internally. A decarbonized Europe can only succeed in a decarbonized world, and for this to happen, a green

Europe must necessarily be a global one too. With climate and the energy transition representing Europe's mission, and with the success or failure of this agenda marking the fate of the European project for decades to come, a green and global Europe must become two sides of the same coin.

Notes

1 Taking historical emissions over the last 100 years that have contributed to the current concentrations of atmospheric CO_2, the EU's share of emissions is higher, at around 15 per cent.

2 Such as China's '1+N' carbon policy or the Biden administration's Build Back Better Framework.

3 In 2003, Mikhail Khodorkovsky, the chief executive of Yukos, a Western-style company with American investors, was arrested and imprisoned in Russia for alleged tax evasion and fraud.

4 For instance, by reaching 40 per cent of renewables in its energy mix, increasing its energy efficiency target to 36–39 per cent by 2030. The package also set out more specific targets to get there, such as ending sales of new petrol and diesel vehicles or planting three billion trees by 2030.

5 During which global average temperatures rose rapidly between 5 and 8°C.

6 I speak of nationalist populists when referring to those political forces that are both nationalist and populist. The former ideologically exalt the concept of the 'nation' and the defence of 'national interests', the latter juxtapose the 'people' to the 'elites'. Not all nationalists are populists, and vice versa. However, often nationalists are also populists, and by mentioning both, I indicate this subset of political forces. These nationalist-populist forces are generally Eurosceptic. This is because nationalists are sceptical of the supranational European project, and populists look at 'Brussels' as the epicentre of a distant 'elite', detached from the 'real' interests of the 'people'.

7 Although with lower than expected renewable production,

demand for coal has also increased, raising further the upward pressure on CO_2 prices.

8 American energy companies, faced with a different political and policy context as far as climate change is concerned, lag well behind their European peers in terms of their commitment to decarbonization.

9 According to the Commission, the effect of the gas price increase on the electricity price is nine times bigger than that of the carbon price increase.

10 Currently these sectors fall under the EU's effort-sharing regulation, i.e. national GDP-based emission targets for sectors which are not included in the ETS.

11 Market tests for the expansion of the Trans Adriatic Pipeline carrying Azerbaijani gas had been largely deserted. It is only against the backdrop of the Ukraine war that interest in the expansion of TAP reawakened.

12 On the different capacities of Middle East and North African countries to navigate the energy transition, see Raimondi and Tagliapietra (2021).

13 'Blue hydrogen' is produced through steam methane reforming, which is used to produce 'grey hydrogen', and is coupled with CCS technologies.

14 'Green hydrogen' refers to the production of hydrogen through electrolysis from renewables-generated electricity.

15 Beyond the 38 bcm that were already agreed, Presidents Putin and Xi agreed during their meeting at the Beijing Winter Olympics in February 2022 to add 10 bcm of natural gas to the pipeline.

16 CBAM prevents the unfair disadvantage to European industry that would be subject to a more ambitious ETS – with phased-out allowances – vis-à-vis third country importers to Europe that would otherwise not. It does not address the potential loss of international competitiveness of European exporters, which would be subject to higher carbon prices than their non-European peers. There could be several solutions to this problem, including export subsidies. But this would probably fall foul of WTO rules. Rather than export subsidies, the EU could address this problem by ramping up its support for green innovation and investment. This could include support for carbon contracts for difference, whereby industries would be guaranteed the coverage of costs in a decarbonization technology when these exceed current carbon prices.

17 For instance, by phasing out free emission allowances.

18 The Energy Community is an international organization that includes the EU and its neighbours, and aims at creating an integrated pan-European energy market.

19 For a broader look at transit migration through Niger, see Stijn and Rijks (2020).

20 This does not imply that US–China competition is not also military in nature, nor that the military dimension – for instance over Taiwan – may not become the area in which systemic competition erupts into open confrontation.

21 Whereas battery recycling rates in Europe are quite high, less than 1 per cent of the critical raw minerals identified by the EU are recovered in this process to date, given the EU still lacks the technological and industrial capacities to make a meaningful step forward in this field.

22 In 2014, as momentum started building behind COP21 in Paris, the EU had announced its plans to achieve a 40 per cent reduction in greenhouse emissions by 2030, a 27 per cent share of renewables in its energy mix, and a 27 per cent improvement in energy efficiency.

23 Whereby the US agreed to cut its greenhouse gas emissions by 26–28 per cent by 2028 compared to 2005 levels and China to cap the increase of its CO_2 emissions by no later than 2030.

24 In the 1990s the EU and the US squabbled over Caribbean bananas. Washington complained about the special trade access granted to Latin American producers to European markets. Tariffs were slapped on by both sides, reaching 100 per cent on select products.

25 Methane is eighty times more potent than CO_2 as a greenhouse gas. However, unlike methane, CO_2 lingers in the atmosphere for centuries rather than only a few decades.

26 At state level, carbon pricing has been adopted in states like Massachusetts and California, and is under consideration or development in several other states on both the east and west coasts.

27 After much wrangling, the Commission proposed a stringent set of rules under which gas and nuclear projects would be eligible under the taxonomy. The rules are particularly strict on gas. New gas power plant projects approved by 2030 would be eligible provided: (a) their direct emissions are lower than 270 g CO_2/kWh or 550 kg CO_2e/kW per year over twenty years; (b) they demonstrate that such energy could not be produced from renewable sources; (c) they replace, without adding more

than 15 per cent capacity, oil or coal plants and by doing so contribute to a reduction of emissions by at least 55 per cent over the power plant's life cycle; (d) they are designed to phase in renewable or low-carbon natural gas by 2035; and (e) they are built in Member States in which electricity is still produced by coal. The technical feasibility of respecting such criteria is questionable, given the levels of energy efficiency and blue hydrogen production implied. Furthermore, the criteria oddly reward those Member States that are yet to make the coal-to-gas switch, penalizing those states that have already made the transition.

28 €60.4 billion is destined for sub-Saharan Africa, out of a €79.5 billion total NDICI budget.

29 With the exception of the UK, which prior to Brexit belonged to the European Emission Trading Scheme, and since 1 January 2021 has established the UK ETS, essentially replicating the same model.

30 With the outbreak of Covid-19, the 2020 data are probably less indicative of overall trends given the sharp reduction of the (online) meetings held, as well as their focus on the pandemic itself.

31 For example, the performance of helicopters in Afghanistan was compromised by the dust and the heat, contributing to the launch of an EU permanent structured cooperation project on 'Helicopter Hot and High Training'.

References

Ahmad, Ali, and Monali Renade. 2021. 'Climate Responsive Economic Recovery: Post-Pandemic Opportunities in Mashreq'. The World Bank. https://openknowledge .worldbank.org/handle/10986/36268.

Ahmed, Bashair, Mahamadou Danda, Audra Grant, and Vassilis Ntousas. 2018. 'The Security–Migration– Development Nexus Revised: A Perspective from the Sahel'. Istituto Affari Internazionali and Foundation for European Progressive Studies. https://www.iai.it/sites /default/files/9788868129729.pdf.

Alcaro, Riccardo, and Nathalie Tocci. 2021. 'Navigating a Covid World: The European Union's Internal Rebirth and External Quest'. *The International Spectator* 56 (2): 1–18. https://doi.org/10.1080/03932729.2021.1911128.

al-Jazeera. 2021. 'DR Congo President Seeks Review of Mining Contracts with China'. *Al-Jazeera*, 11 September, sec. News. https://www.aljazeera.com/news/2021/9/11/dr -congo-leader-seeks-review-of-mining-deals-with-china.

Allison, Graham T. 2017. *Destined for War: Can America and China Escape Thucydides's Trap?* Boston, MA: Houghton Mifflin Harcourt.

Allison, Graham T. 2021. 'Will America's Green Future Be Red?' *Boston Globe*, 29 October, sec. Opinion. https:// www.bostonglobe.com/2021/10/29/opinion/will-americas -green-future-be-red/.

Andrijevic, Marina, and Joe Ware. 2021. 'Lost & Damaged:

A Study of the Economic Impact of Climate Change on Vulnerable Countries'. Christian Aid & ActAlliance. https://reliefweb.int/sites/reliefweb.int/files/resources/Lost _and_Damaged_-_A_study_of_the_economic_impact_of _climate_change_on_vulnerable_countries.pdf.

Archer, David. 2016. *The Long Thaw: How Humans Are Changing the Next 100,000 Years of Earth's Climate*. Princeton, NJ: Princeton University Press.

Aydıntaşbaş, Asli, and Susi Dennison. 2021. 'New Energies: How the European Green Deal Can Save the EU's Relationship with Turkey'. Policy Brief. European Council on Foreign Relations. https://ecfr.eu/publication/new -energies-how-the-european-green-deal-can-save-the-eus -relationship-with-turkey/.

Babiš, Andrej. 2021a. 'It Is Absolutely Crucial for Individual States to Choose Their Own Energy Mix to Achieve Carbon Neutrality'. Government of the Czech Republic. https://www.vlada.cz/en/clenove-vlady/premier/speeches /andrej-babis-it-is-absolutely-crucial-for-individual-states -to-choose-their-own-energy-mix-to-achieve-carbon -neutrality-191508/.

Babiš, Andrej. 2021b. 'I'll Fight for you Until My Body Is Torn Apart!' Facebook Post. https://www.anobudelip .cz/novydopis/?fbclid=IwAR0hPr9J0K1_L6vgKaPTI0 -g8vokpvRqRKmRIso_UEMY-Cxy9VuEqbF2okU.

Baconi, Tareq. 2017. 'Pipelines and Pipedreams: How the EU Can Support a Regional Gas Hub in the Eastern Mediterranean'. Policy Brief. European Council on Foreign Relations. https://ecfr.eu/wp-content/uploads/ECFR211 _-_PIPELINES_AND_PIPEDREAMS.pdf.

Baer, Dan. 2021. 'Tracking Biden's Progress on a Foreign Policy for the Middle Class'. *Carnegie Endowment for International Peace*, 6 April. https://carnegieendowment .org/2021/04/06/tracking-biden-s-progress-on-foreign -policy-for-middle-class-pub-84236.

Bayer, Lili. 2019. 'Meet von der Leyen's "Geopolitical Commission"'. *Politico*, 12 April. https:// www.politico.eu/article/meet-ursula-von-der-leyen- geopolitical-commission/.

Bergamaschi, Luca, Nick Mabey, Jonathan Gaventa, and Camilla Born. 2016. 'EU Foreign Policy in a Changing Climate: A Climate and Energy Strategy for Europe's Long-Term Security'. Discussion Draft. E3G Think Tank. https://www.e3g.org/wp-content/uploads/E3G_EU_foreign_policy_energy_climate.pdf.

Betz, Hans-Georg. 2020. 'Coronavirus-19's Victims: Populism'. In Tamir Bar-On and Barbara Molas (eds) *Responses to the COVID-19 Pandemic by the Radical Right: Scapegoating, Conspiracy Theories and New Narratives*. Stuttgart: Ibidem Verlag, pp. 45–53.

Bianchi, Margerita, and Lorenzo Colantoni. 2021. 'Green Deal Watch No. 5'. Istituto Affari Internazionali. https://www.iai.it/en/pubblicazioni/green-deal-watch-no-5.

Bielkova, Olga. 2021. 'Ukraine's Energy Future Is Tied to European Integration'. *Atlantic Council*, 6 March. https://www.atlanticcouncil.org/blogs/ukrainealert/ukraines-energy-future-is-tied-to-european-integration/.

Boas, Morten, and Francesco Strazzari. 2020. 'Governance, Fragility and Insurgency in the Sahel: A Hybrid Political Order in the Making'. *The International Spectator 55* (4): 1–17. https://doi.org/10.1080/03932729.2020.1835324.

Bobba, Giuliano, and Nicolas Hubé. 2021. 'Populism and Covid-19 in Europe: What We Learned from the First Wave of the Pandemic'. *LSE Blog*. 14 April. https://blogs.lse.ac.uk/europpblog/2021/04/20/populism-and-covid-19-in-europe-what-we-learned-from-the-first-wave-of-the-pandemic/.

Bond, Ian. 2021. 'Why Have Europe's Energy Prices Spiked and What Can the EU Do about Them?' Centre for European Reform. 28 October. https://www.cer.eu/insights/why-have-europes-energy-prices-spiked.

Bordoff, Jason. 2020. 'It's Time to Put Climate Action at the Center of US Foreign Policy'. *Foreign Policy*, 27 July, sec. Voice. https://foreignpolicy.com/2020/07/27/climate-change-foreign-policy/.

Bordoff, Jason, and Meghan O'Sullivan. 2021. 'Green Upheaval: The New Geopolitics of Energy'. *Foreign Affairs*,

January/February. https://www.foreignaffairs.com/articles /world/2021-11-30/geopolitics-energy-green-upheaval.

Bos, Julie, Lorena Gonzalez, and Joe Thwaites. 2021. 'Are Countries Providing Enough to the $100 Billion Climate Finance Goal?' *World Resources Institute*, 7 October, sec. Finance. https://www.wri.org/insights/developed-countries -contributions-climate-finance-goal.

Bradford, Anu. 2020. *The Brussels Effect: How the European Union Rules the World*. Oxford: Oxford University Press. https://doi.org/10.1093/oso/9780190088583.001.0001.

Brands, Hal. 2021. 'The Emerging Biden Doctrine'. *Foreign Affairs*, 29 June. https://www.foreignaffairs.com/articles /united-states/2021-06-29/emerging-biden-doctrine.

Carty, Tracy, Jan Kowalzig, and Bertram Zagema. 2020. 'Climate Finance Shadow Report 2020. Assessing Progress Towards the $100 Billion Commitment'. Oxfam. https:// oxfamilibrary.openrepository.com/bitstream/handle /10546/621066/bp-climate-finance-shadow-report-2020 -201020-en.pdf.

Ciravegna, Luciano, and Snejina Michailova. 2022. 'Why the World Economy Needs, but Will Not Get, More Globalization in the Post-Covid-19 Decade'. *Journal of International Business Studies* 53, 172–86. https://doi.org /10.1057/s41267-021-00467-6.

Climate Action Network. 2021. 'Consultation on the Regional Multi-Annual Indicative Program for Sub-Saharan Africa (2021–2027)'. Climate Action Network Europe. https:// caneurope.org/content/uploads/2021/04/Green-Transition -recommendations-for-the-EU-Sub-Saharan-Africa -regional-MIP_March2021-.pdf.

Collomb, Jean-Daniel. 2014. 'The Ideology of Climate Change Denial in the United States'. *European Journal of American Studies* 9 (1). https://doi.org/10.4000/ejas .10305.

Cook, Steven A. 2017. 'Turkey: Friend or Frenemy? A Tangled Relationship Keeps Getting Worse'. *Council on Foreign Relations*, 13 November. https://www.cfr.org /blog/turkey-friend-or-frenemy-tangled-relationship-keeps -getting-worse.

Cooley, Alexander, and Daniel Nexon. 2021. 'The Illiberal Tide: Why International Order Is Tilting towards Autocracy'. In Charles A. Kupchan and Leslie Vinjamuri (eds) *Anchoring the World*. Foreign Affairs, pp. 51–67. https://www.foreignaffairs.com/system/files/pdf/2021/FA _Anchoring%20the%20World_digital.pdf.

COP26. 2021. 'Mobilise Finance'. *UN Climate Change Conference UK 2021*. https://ukcop26.org/cop26-goals /finance/.

Cowles, Maria G., and Michelle Egan. 2012. 'The Evolution of the Transatlantic Partnership'. Working Paper 3. Transworld. http://transworld.iai.it/wp-content/uploads /2012/10/TW_WP_03.pdf.

Cui, Ryna Yiyun, Nathan Hultman, Diyang Cui, et al. 2021. 'A Plant-by-Plant Strategy for High-Ambition Coal Power Phaseout in China'. *Nature Communications* 12 (1): 1468. https://doi.org/10.1038/s41467-021-21786-0.

Cullenward, Danny, and David G. Victor. 2020. *Making Climate Policy Work*. Cambridge: Polity.

Damro, Chad. 2012. 'Market Power Europe'. *Journal of European Public Policy* 19 (5): 682–99. https://doi.org/10 .1080/13501763.2011.646779.

Davis Cross, Mai'a K. 2018. 'Partners at Paris? Climate Negotiations and Transatlantic Relations'. *Journal of European Integration* 40 (5): 571–86. https://doi.org/10 .1080/07036337.2018.1487962.

Dempsey, Judy. 2018. 'Judy Asks: Is Europe Afraid of Migration?' *Carnegie Europe*, 13 September. https:// carnegieeurope.eu/strategiceurope/77246.

Derviş, Kemal, and Nathalie Tocci. 2022. 'Liberal Democratic Values and the Future of Multilateral Cooperation'. In Brahima S. Coulibaly and Kemal Derviş (eds) *Essays on a 21st Century Multilateralism that Works for All*. Washington, DC: Brookings Institution, pp. 57–65. https:// www.brookings.edu/wp-content/uploads/2022/02/21st -Century-Multilateralism.pdf.

Di Bartolomeo, Anna, Thibaut Jaulin, and Delphine Perrin. 2011. 'Migration Profile – Niger'. Consortium for Applied Research on International Migration (CARIM). European

University Institute. https://cadmus.eui.eu/bitstream /handle/1814/22442/migration%20profile%20EN %20Niger%20-%20links.pdf?sequence=1&is Allowed=y.

Diamond, Larry. 2015. 'Facing Up to the Democratic Recession'. *Journal of Democracy* 26 (1): 141–55.

Dokos, Thanos, Eleonora Poli, Chiara Rosselli, Eduard Soler i Lecha, and Nathalie Tocci. 2013. 'Eurocriticism: The Eurozone Crisis and Anti-Establishment Groups in Southern Europe'. IAI Working Papers 13 | 33. Istituto Affari Internazionali. https://ciaotest.cc.columbia.edu/wps /iai/0030636/f_0030636_24777.pdf.

Drezner, Daniel W. 2021. 'Real Talk about a Foreign Policy for the Middle Class'. *The Washington Post*, 20 May. https://www.washingtonpost.com/outlook/2021/05/20 /real-talk-about-foreign-policy-middle-class/.

DW. 2021. 'Natural Disasters Cost $280 Billion in 2021: German Insurance Firm'. *Deutsche Welle*, 1 October. https://www.dw.com/en/natural-disasters-cost-280-billion -in-2021-german-insurance-firm/a-60378575?maca=en -rss-en-all-1573-rdf.

EBA. 2021. 'Building a European Battery Industry'. *InnoEnergy – European Battery Alliance*. https://www .eba250.com/?cn-reloaded=1.

EEAS. 2011. 'Strategy for Security and Development in the Sahel'. European External Action Service. https:// eeas.europa.eu/archives/docs/africa/docs/sahel_strategy_en .pdf.

EEAS. 2016. 'Shared Vision, Common Action: A Stronger Europe. A Global Strategy for the European Union's Foreign and Security Policy'. European External Action Service. https://eeas.europa.eu/sites/default/files/eugs _review_web_0.pdf.

EEAS. 2021. 'The European Union's Integrated Strategy in the Sahel'. European External Action Service & General Secretariat of the Council. https://data.consilium.europa .eu/doc/document/ST-7723-2021-INIT/en/pdf.

EIB. 2021a. *The EIB Climate Survey: The Climate Crisis in a Covid 19 World: Calls for a Green Recovery.*

European Investment Bank. https://data.europa.eu/doi/10.2867/5219.

EIB. 2021b. *EIB Investment Report 2020/2021: Building a Smart and Green Europe in the COVID 19 Era*. European Investment Bank. https://data.europa.eu/doi/10.2867/904099.

Eicke, Laima, Silvia Weko, Maria Apergi, and Adela Marian. 2021. 'Pulling Up the Carbon Ladder? Decarbonization, Dependence, and Third-Country Risks from the European Carbon Border Adjustment Mechanism'. *Energy Research & Social Science* 80: 102240. https://doi.org/10.1016/j.erss.2021.102240.

Elagina, D. 2021. 'Share of Gas Supply from Russia in Europe in 2020, by Selected Country'. *Statista*, 6 September. https://www.statista.com/statistics/1201743/russian-gas-dependence-in-europe-by-country/.

Erickson, Andrew S., and Gabriel Collins. 2021. 'Competition with China Can Save the Planet: Pressure, Not Partnership, Will Spur Progress on Climate Change'. *Foreign Affairs*, May/June. https://www.foreignaffairs.com/articles/united-states/2021-04-13/competition-china-can-save-planet.

Erzini Vernoit, Iskander, Sonia Dunlop, James Hawkins, and Dileimy Orozco. 2020. 'The European Investment Bank: Becoming the EU Climate Bank'. E3G Think Tank. https://www.e3g.org/wp-content/uploads/09_07_20_E3G-EIB-Becoming-EU-Climate-Bank-report.pdf.

Euractiv. 2021. 'EU Lawmakers Give Final Approval to Bloc's Green Transition Fund'. *Euractiv*, 19 May. https://www.euractiv.com/section/energy-environment/news/eu-lawmakers-give-final-approval-to-blocs-green-transition-fund/.

European Climate Foundation. 2019. 'Majority of Voters Want Political Parties to Tackle Global Warming'. European Climate Foundation, 16 April. https://europeanclimate.org/resources/majority-of-voters-want-political-parties-to-tackle-global-warming/.

European Commission. 2019. 'Communication from the Commission to the European Parliament, the European

Council, the Council, the European Economic and Social Committee and the Committee of the Regions the European Green Deal'. COM(2019) 640 final. European Commission. https://eur-lex.europa.eu/legal-content /EN/TXT/?qid=1576150542719&uri=COM%3A2019 %3A640%3AFIN.

European Commission. 2020a. 'Questions and Answers – the EU Budget for External Action in the Next Multiannual Financial Framework'. Questions and Answers. European Commission. https://ec.europa.eu/neighbourhood -enlargement/news/questions-and-answers-eu-budget -external-action-next-multiannual-financial-framework -2020-06_en.

European Commission. 2020b. 'Communication from the Commission to the European Parliament, the Council, the European Economic and Social Committee and the Committee of the Regions "A Hydrogen Strategy for a Climate-Neutral Europe"'. COM(2020) 301 final. European Commission. https://ec.europa.eu/energy/sites /ener/files/hydrogen_strategy.pdf.

European Commission. 2020c. 'Communication from the Commission to the European Parliament, the Council, the European Economic and Social Committee and the Committee of the Regions "An EU Strategy to Harness the Potential of Offshore Renewable Energy for a Climate Neutral Future"'. SWD(2020) 273 final. European Commission. https://eur-lex.europa.eu/legal-content/EN /TXT/PDF/?uri=CELEX:52020DC0741&from=EN.

European Commission. 2020d. 'White Paper – Levelling the Playing Field as Regards Foreign Subsidies'. COM(2020) 253 final. European Commission. https:// ec.europa.eu/competition/international/overview/foreign_ subsidies_white_paper.pdf.

European Commission. 2020e. 'Joint Communication to the European Parliament, the European Council and the Council: "A New EU–US Agenda for Global Change"'. JOIN(2020) 22 final. European Commission. https://ec .europa.eu/info/sites/default/files/joint-communication-eu -us-agenda_en.pdf.

European Commission. 2021a. 'Questions and Answers – Making Our Energy System Fit for Our Climate Targets'. Questions and Answers. European Commission. https://ec .europa.eu/commission/presscorner/detail/en/QANDA_21 _3544.

European Commission. 2021b. 'Questions and Answers – Commission Communication on Energy Prices'. Questions and Answers. European Commission. https://ec.europa.eu /commission/presscorner/detail/en/qanda_21_5202.

European Commission. 2021c. 'Communication from the Commission to the European Parliament, the Council, the European Economic and Social Committee and the Committee of the Regions "Fit for 55": Delivering the EU's 2030 Climate Target on the Way to Climate Neutrality'. COM(2021) 550 final. European Commission. https://eur -lex.europa.eu/legal-content/EN/ALL/?uri=COM:2021: 550:FIN.

European Commission. 2021d. 'Modernisation Fund'. European Commission. https://ec.europa.eu/clima/eu -action/funding-climate-action/modernisation-fund_en.

European Commission. 2021e. 'Communication from the Commission to the European Parliament, the Council, the European Economic and Social Committee and the Committee of the Regions "Forging a Climate-Resilient Europe – the New EU Strategy on Adaptation to Climate Change"'. COM(2021) 82 final. https://eur-lex.europa.eu /legal-content/EN/TXT/?uri=COM:2021:82:FIN.

European Commission. 2021f. 'Initiative for Coal Regions in Transition in the Western Balkans and Ukraine'. European Commission. https://ec.europa.eu/energy/topics /oil-gas-and-coal/coal-regions-in-the-western-balkans-and -ukraine/initiative-coal-regions-transition-western-balkans -and-ukraine_en.

European Commission. 2021g. 'Communication from the Commission to the European Parliament, the Council, the European Economic and Social Committee and the Committee of the Regions "Updating the 2020 New Industrial Strategy: Building a Stronger Single Market for Europe's Recovery"'. COM(2021) 350 final. European

Commission. https://ec.europa.eu/info/sites/default/files
/communication-industrial-strategy-update-2020_en.pdf.

European Commission. 2021h. 'Countries and Regions –
Turkey'. Questions and Answers. European Commission.
https://ec.europa.eu/trade/policy/countries-and-regions
/countries/turkey/.

European Commission. 2021i. '2021 State of the Union
Address by President von der Leyen'. Speech. European
Commission. https://ec.europa.eu/commission/presscorner
/detail/en/SPEECH_21_4701.

European Commission. 2021j. 'Ahead of G20 Summit and
COP26, President von der Leyen Sets Out EU Priorities'.
European Commission. https://ec.europa.eu/commission
/presscorner/detail/en/AC_21_5675.

European Council. 2003. 'European Security Strategy'.
15895/03, PESC 787. Council of the European Union.
https://data.consilium.europa.eu/doc/document/ST-15895
-2003-INIT/en/pdf.

European Council. 2008. 'Climate Change and International
Security. Paper from the High Representative and the
European Commission to the European Council'. S113/08.
Council of the European Union. https://www.consilium
.europa.eu/uedocs/cms_data/docs/pressdata/en/reports
/99387.pdf.

European Council. 2012. 'Towards a Genuine Economic and
Monetary Union. Report by President of the European
Council Herman Van Rompuy'. EUCO 120/12 PRESSE
296. Council of the European Union. https://www
.consilium.europa.eu/media/21570/131201.pdf.

European Council. 2020. 'Special Meeting of the European
Council (1 and 2 October 2020) – Conclusions'. EUCO
13/20. Council of the European Union. https://www
.consilium.europa.eu/media/45910/021020-euco-final
-conclusions.pdf.

European Council. 2021a. 'Council Conclusions on Climate
and Energy Diplomacy – Delivering on the External
Dimension of the European Green Deal'. 5263/21. Council
of the European Union. https://www.consilium.europa.eu
/media/48057/st05263-en21.pdf.

European Council. 2021b. 'Press Release 29102021 – EU Council Approves 2020 Climate Finance Figure'. EUCO 120/12 PRESSE 296. Council of the European Union. https://www.consilium.europa.eu/en/press/press-releases/2021/10/29/council-approves-2020-climate-finance-figure/.

European Energy Poverty Advisory Hub. 2021. 'Member State Reports on Energy Poverty 2019'. European Commission. https://energy-poverty.ec.europa.eu/discover/practices-and-policies-toolkit/publications/epov-member-state-reports-energy-poverty-2019_en.

European Parliament. 2021. 'Legislative Train Schedule'. European Parliament. https://www.europarl.europa.eu/legislative-train/theme-regional-development-regi/file-just-transition-fund.

Eurostat. 2021. 'EU Imports of Energy Products – Recent Developments'. Eurostat. https://ec.europa.eu/eurostat/statistics-explained/index.php?title=EU_imports_of_energy_products_-_recent_developments#Main_suppliers_of_natural_gas_and_petroleum_oils_to_the_EU.

Evans, Simon, Josh Gabbatiss, Robert McSweeney, and Aruna Chandrasekhar. 2021. 'COP26: Key Outcomes Agreed at the UN Climate Talks in Glasgow'. *CarbonBrief*, 15 November. https://www.carbonbrief.org/cop26-key-outcomes-agreed-at-the-un-climate-talks-in-glasgow.

Eyl-Mazzega, Marc-Antoine, and Carole Mathieu. 2020. 'The European Union and the Energy Transition'. In Manfred Hafner and Simone Tagliapietra (eds) *The Geopolitics of the Global Energy Transition* (Lecture Notes in Energy 73). Cham: Springer International Publishing, pp. 27–46.

Falchetta, Giacomo, and Michel Noussan. 2021. 'Electric Vehicle Charging Network in Europe: An Accessibility and Deployment Trends Analysis'. *Transportation Research Part D: Transport and Environment* 94: 102813. https://doi.org/10.1016/j.trd.2021.102813.

Farand, Chloé. 2020. 'EIB Approves €1 Trillion Green Investment Plan to Become "Climate Bank"'. *Climate Home News*, 12 November. https://www.climatechangenews.com

/2020/11/12/eib-approves-e1-trillion-green-investment
-plan-become-climate-bank/.

Faria, Fernanda. 2004. 'La Gestion des Crises en Afrique
Subsaharienne: Le Rôle de l'Union Européenne'.
Occasional Paper 55. EU Institute for Security Studies.
https://www.iss.europa.eu/content/la-gestion-des-crises-en
-afrique-subsaharienne-le-rôle-de-lunion-européenne.

Farrell, Henry, and Abraham L. Newman. 2019. 'Weaponized
Interdependence: How Global Economic Networks Shape
State Coercion'. *International Security* 44 (1): 42–79.
https://doi.org/10.1162/isec_a_00351.

Fieschi, Catherine. 2022. 'The Green Wedge Essays.
Climate Views in Context'. European Policy Institute
– Counterpoint. https://counterpoint.uk.com/wp-content
/uploads/2021/05/Green-Wedge-Essays-Counterpoint.pdf.

Fleeson, William. 2021. 'Turkey Could Face Over $900 Million
in CBAM-Related Export Costs: Report'. *IHS Markit*, 8
December. https://ihsmarkit.com/research-analysis/turkey
-could-face-over-900-million-in-cbamrelated-export-costs
.html.

Fleming, Sam, and Chris Giles. 2021. 'OECD Seeks Global
Plan for Carbon Prices to Avoid Trade Wars'. *The Financial
Times*, 13 September. https://www.ft.com/content
/334cf17a-e1f1-4837-807a-c4965fe497f3.

Franza, Luca. 2021. 'Clean Molecules across the
Mediterranean: The Potential for North African Hydrogen
Imports into Italy and the EU'. Istituto Affari Internazionali.
https://www.iai.it/sites/default/files/9788893681834.pdf.

Franza, Luca, Margherita Bianchi, and Luca Bergamaschi.
2020. 'Geopolitics and Italian Foreign Policy in the Age of
Renewable Energy'. IAI Working Papers 20 | 13. Istituto
Affari Internazionali. https://www.iai.it/sites/default/files
/iaip2013.pdf.

Funk, Cary, and Meg Hefferson. 2019. 'US Public Views on
Climate and Energy'. Pew Research Center. https://www
.pewresearch.org/science/2019/11/25/u-s-public-views-on
-climate-and-energy/.

Gasmi, Farid, and Imène Laourari. 2017. 'Has Algeria Suffered
from the Dutch Disease? Evidence from 1960–2016 Data'.

Revue d'économie politique 127 (6): 1029–58. https://doi .org/10.3917/redp.276.1029.

Gazprom. 2021. 'Gazprom and Mongolian Government Discuss Progress of Feasibility Study for Soyuz Vostok Gas Pipeline Project'. 22 October. https://www.gazprom.com /press/news/2021/october/article540773/.

GCCA+. 2021. 'What Is the EU GCCA+ Initiative?' *The Global Climate Change Alliance Plus Initiative.* https:// www.gcca.eu/about-eu-gcca.

Goldthau, Andreas. 2017. 'The G20 Must Govern the Shift to Low-Carbon Energy'. *Nature* 546 (7657): 203–5. https:// doi.org/10.1038/546203a.

Goldthau, Andreas, and Kirsten Westphal. 2019. 'Why the Global Energy Transition Does Not Mean the End of the Petrostate'. *Global Policy* 10 (2): 279–83. https://doi.org /10.1111/1758-5899.12649.

Gotev, Georgi. 2021. 'Putin Blames EU Green Policies for Energy Price Spike'. *Euractiv,* 14 October. https://www .euractiv.com/section/global-europe/news/putin-blames-eu -green-policies-for-the-energy-price-spike/.

Goxho, Delina, and Louis Mourier. 2021. 'In the Sahel, Europe Needs to Rethink Its Approach to Climate Change and Conflict'. *Voxeurop,* 6 January. https://voxeurop.eu /en/in-the-sahel-europe-needs-to-rethink-its-approach-to -climate-change-and-conflict/.

Grübler, Julia, Roman Stöllinger, and Gabriele Tondl. 2021. 'Are EU Trade Agreements in Line with the European Green Deal?' Vienna Institute for International Economic Studies (WiiW), 15 February. https://wiiw.ac.at/are-eu -trade-agreements-in-line-with-the-european-green-deal -n-484.html.

Gulley, Andrew L., Erin A. McCullough, and Kim B. Shedd. 2019. 'China's Domestic and Foreign Influence in the Global Cobalt Supply Chain'. *Resources Policy* 62: 317–23. https://doi.org/10.1016/j.resourpol.2019.03.015.

Hache, Emmanuel. 2018. 'Do Renewable Energies Improve Energy Security in the Long Run?' *International Economics* 156: 127–35. https://doi.org/10.1016/j.inteco.2018.01 .005.

Hafner, Manfred, and Simone Tagliapietra (eds) 2020. *The Geopolitics of the Global Energy Transition* (Lecture Notes in Energy 73). Cham: Springer Open.

Hafner, Manfred, and Alessa Wochner. 2020. 'How Tectonic Shifts in Global Energy Are Affecting Global Governance'. In Leonid Grigoryev and Adrian Pabst (eds) *Global Governance in Transformation*. Cham: Springer International Publishing, pp. 147–62.

Hall, Peter A., and David Soskice. 2001. *Varieties of Capitalism*. Oxford: Oxford University Press.

Hass, Ryan. 2021. 'How China Is Responding to Escalating Strategic Competition with the US'. *Brookings Institution*, 1 March. https://www.brookings.edu/articles/how-china-is-responding-to-escalating-strategic-competition-with-the-us/.

Hernández-Morales, Aitor, Karl Mathiesen, Stuart Lau, and Giorgio Leali. 2021. 'Fears over China's Muslim Forced Labor Loom over EU Solar Power'. *Politico*, 10 February. https://www.politico.eu/article/xinjiang-china-polysilicon-solar-energy-europe/.

Hook, Leslie, Neil Hume, and Jim Pickard. 2021. 'COP26 Pact to End Coal Use Undermined as US Fails to Sign'. *The Financial Times*, 11 April. https://www.ft.com/content/94584f35-350f-4391-894e-5af2023eb9ab.

Hungary Today. 2021. 'PM Orban: EU Needs to Rethink Policy on Energy Prices'. *Hungary Today*, 6 October. https://hungarytoday.hu/orban-energy-prices-eu-policy/.

IEA. 2018. *World Energy Outlook 2018*. World Energy Outlook. OECD International Energy Agency. https://iea.blob.core.windows.net/assets/77ecf96c-5f4b-4d0d-9d93-d81b938217cb/World_Energy_Outlook_2018.pdf.

IEA. 2019. *World Energy Outlook 2019*. World Energy Outlook. OECD International Energy Agency. https://doi.org/10.1787/caf32f3b-en.

IEA. 2021a. 'Net Zero by 2050: A Roadmap for the Global Energy Sector'. OECD International Energy Agency. https://doi.org/10.1787/c8328405-en.

IEA. 2021b. 'Statement on Recent Developments in Natural Gas and Electricity Markets (21 September 2021)'. *IEA*

News   (OECD).   https://www.iea.org/news/statement
-on-recent-developments-in-natural-gas-and-electricity
-markets.

IEA. 2021c. *World Energy Outlook 2021*. World Energy
Outlook. OECD International Energy Agency. https://doi
.org/10.1787/14fcb638-en.

IEA. 2021d. 'Financing Clean Energy Transitions in Emerging
and   Developing   Economies'.   OECD   International
Energy   Agency.   https://www.iea.org/reports/financing
-clean-energy-transitions-in-emerging-and-developing
-economies.

Institut Montaigne. 2019. 'Les Gilets Jaunes: la partie
émergée de la crise sociale française?' *Blog de l'Institut
Montaigne*, 20 March. http://www.institutmontaigne.org
/blog/les-gilets-jaunes-la-partie-emergee-de-la-crise-sociale
-francaise.

International Organization for Migration. 2020. 'Niger:
Flow   Monitoring   Report   37   (July   2020)'.   United
Nations.   https://migration.iom.int/reports/niger-—-flow
-monitoring-report-37-july-2020.

IPCC. 2021. 'Climate Change Widespread, Rapid, and
Intensifying'. International Panel on Climate Change.
https://www.ipcc.ch/2021/08/09/ar6-wg1-20210809-pr/.

Ipsos. 2021. '#ClimateOfChange. Pan-European Survey.
Main Multi-Country Report'. Belgium: Ipsos. https://
mk0eeborgicuypctuf7e.kinstacdn.com/wp-content
/uploads/2021/04/IPSOS-Multi-Country-Report-complete
.FINAL_.pdf.

IRENA. 2019a. 'A New World: The Geopolitics of the Energy
Transformation'. International Renewable Energy Agency.
https://www.irena.org/-/media/Files/IRENA/Agency
/Publication/2019/Jan/Global_commission_geopolitics
_new_world_2019.pdf.

IRENA. 2019b. 'Global Energy Transformation: A Roadmap
to 2050'. International Renewable Energy Agency. https://
www.irena.org/-/media/Files/IRENA/Agency/Publication
/2019/Apr/IRENA_Global_Energy_Transformation_2019
.pdf.

Jaganmohan, Madhumitha. 2021. 'Global Market Share of

the World's Leading Wind Turbine Manufacturers in 2018, Based on Sales'. *Statista*. 6 October. https://www.statista .com/statistics/272813/market-share-of-the-leading-wind -turbine-manufacturers-worldwide/.

Jones, Liam. 2021. 'Europe Reaches $500bn in Green Investment – Climate Bonds Market Intel Reports'. *Climate Bonds Initiative*, 17 May. https://www.climatebonds.net /2021/05/europe-reaches-500bn-green-investment-climate -bonds-market-intel-reports.

Jones, Sam, James Shotter, and Guy Chazan. 2021. 'Covid Backlash: Europe's Populists Eye Opportunity in Never-Ending Pandemic'. *The Financial Times*, 1 December. https://www.ft.com/content/7ef50a97-c12d-4905-b6da -75c3c7bb4f16.

JRC and European Commission. 2020. *Critical Raw Materials for Strategic Technologies and Sectors in the EU: A Foresight Study*. European Commission & Joint Research Centre. https://data.europa.eu/doi/10.2873/58081.

Kaplan, Thomas, Chris Buckley, and Brad Plumer. 2021. 'US Bans Imports of Some Chinese Solar Materials Tied to Forced Labor'. *The New York Times*, 24 June. https:// www.nytimes.com/2021/06/24/business/economy/china -forced-labor-solar.html.

Kim, Patricia M. 2021. 'Working towards Responsible Competition in China'. *Brookings Institution*, 10 August. https://www.brookings.edu/blog/order-from-chaos/2021 /10/08/working-toward-responsible-competition-with -china/.

Kleimann, David, and William Eacho. 2021. 'Who Is Afraid of the EU's Carbon Border Adjustment Plan?' *The Hill*, 13 October. https://thehill.com/opinion/energy -environment/576637-who-is-afraid-of-the-eus-carbon -border-adjustment-plan.

Krane, Jim, and Robert Idel. 2021. 'More Transitions, Less Risk: How Renewable Energy Reduces Risks from Mining, Trade and Political Dependence'. *Energy Research & Social Science* 82: 102311. https://doi.org/10.1016/j.erss .2021.102311.

La Fabrique de la Cité. 2020. 'Les Gilets Jaunes: simple

révolte anti-métropolitaine ou symptôme d'une crise plus profonde?' *Blog La Fabrique de la Cité*, 10 March. https://www.lafabriquedelacite.com/publications/les-gilets-jaunes-simple-revolte-anti-metropolitaine-ou-symptome-dune-crise-plus-profonde/.

Lam, Long T., Lee Branstetter, and Inês M.L. Azevedo. 2017. 'China's Wind Industry: Leading in Deployment, Lagging in Innovation'. *Energy Policy* 106: 588–99. https://doi.org/10.1016/j.enpol.2017.03.023.

Lavenex, Sandra. 2004. 'EU External Governance in "Wider Europe"'. *Journal of European Public Policy* 11 (4): 680–700. https://doi.org/10.1080/1350176042000248098.

Lebovich, Andrew. 2021. 'After Barkhane: What France's Military Drawdown Means for the Sahel'. *European Council on Foreign Relations*, 2 August. https://ecfr.eu/article/after-barkhane-what-frances-military-drawdown-means-for-the-sahel/.

Legrain, Philippe. 2020. 'The Coronavirus Is Killing Globalization as We Know It'. *Foreign Policy*, 12 March. https://foreignpolicy.com/2020/03/12/coronavirus-killing-globalization-nationalism-protectionism-trump/.

Leonard, Mark, Jean Pisani-Ferry, Jeremy Shapiro, Simone Tagliapietra, and Guntram Wolff. 2021. 'The Geopolitics of the European Green Deal'. Policy Brief. European Council on Foreign Relations. https://ecfr.eu/publication/the-geopolitics-of-the-european-green-deal/.

Lockwood, Matthew. 2018. 'Right-Wing Populism and the Climate Change Agenda: Exploring the Linkages'. *Environmental Politics* 27 (4): 712–32. https://doi.org/10.1080/09644016.2018.1458411.

Loveluck, Louisa, and Mustafa Salim. 2021. 'Climate Change in Iraq Poisons Fertile Crescent Farmland, Empties Villages'. *The Washington Post*, 21 October. https://www.washingtonpost.com/world/interactive/2021/iraq-climate-change-tigris-euphrates/.

Lucarelli, Sonia, and Ian Manners (eds) 2006. *Values and Principles in European Union Foreign Policy* (Routledge Advances in European Politics 37). London: Routledge.

Luciani, Giacomo. 2020. 'The Impacts of the Energy Transition on Growth and Income Distribution'. In Manfred Hafner and Simone Tagliapietra (eds) *The Geopolitics of the Global Energy Transition* (Lecture Notes in Energy 73). Cham: Springer International Publishing, pp. 305–19.

Manners, Ian. 2002. 'Normative Power Europe: A Contradiction in Terms?' *Journal of Common Market Studies* 40 (2): 235–58. https://doi.org/10.1111/1468-5965.00353.

Mathiesen, Karl, and Zack Colman. 2021. '4 Sore Points Between the US and EU on Climate'. *Politico*, 15 June. https://www.politico.eu/article/eu-us-climate-arguments/.

Mélenchon, Jean-Luc [La France Insoumise]. 2019. 'Le vert comme alibi'. *Jean-Luc Mélenchon le blog*, 15 December. https://melenchon.fr/2019/12/15/le-vert-comme-alibi/.

Mesopotamia Revitalization Initiative. 2021. 'Mesopotamia Revitalization Project. A Climate Change Initiative to Transform Iraq and the Middle East'. Iraqi Presidency. https://presidency.iq/EN/Details.aspx?id=3437.

Mitter, Rana. 2021. 'The World China Wants: How Power Will – and Won't – Reshape Chinese Ambitions'. *Foreign Affairs*, January. https://www.foreignaffairs.com/articles/china/2020-12-08/world-china-wants.

Morningstar, Richard L., Andras Simonyi, Olga Khakova, and Irina Markina. 2020. 'European Energy Diversification: How Alternative Sources, Routes, and Clean Technologies Can Bolster Energy Security and Decarbonization'. *Atlantic Council*, 1 September. https://www.atlanticcouncil.org/in-depth-research-reports/issue-brief/european-energy-diversification-how-alternative-sources-and-routes-can-bolster-energy-security-and-decarbonization/.

Morrison, Lee. 2017. 'Southern Gas Corridor: The Geopolitical and Geo-Economic Implications of an Energy Mega-Project'. *The Journal of Energy and Development* 43 (1/2): 251–91.

Murtaugh, Dan. 2021. 'Why It's So Hard for the Solar Industry to Quit Xinjiang'. *Bloomberg*, 19 February, sec. Bloomberg Green. https://www.bloomberg.com

/news/articles/2021-02-10/why-it-s-so-hard-for-the-solar -industry-to-quit-xinjiang.

National Rally [Rassemblement National]. 2020. 'La Délégation Française RN du Groupe Identité et Démocratie au Parlement Européen Vient de s'Abstenir lors du Vote sur le Pacte Vert'. *Blog RN*, 15 January. https:// rassemblementnational.fr/communiques/la-delegation -francaise-rn-du-groupe-identite-et-democratie-au -parlement-europeen-vient-de-sabstenir-lors-du-vote-sur -le-pacte-vert/.

Neill, D.A., and E. Speed. 2012. 'The Strategic Implications of China's Dominance of the Global Rare Earth Elements (REE) Market'. Centre for Operational Research and Analysis.

Nicholas, Thomas, Galen Hall, and Colleen Schmidt. 2020. 'The Faulty Science, Doomism, and Flawed Conclusions of Deep Adaptation'. *OpenDemocracy*, 14 July. https:// www.opendemocracy.net/en/oureconomy/faulty-science -doomism-and-flawed-conclusions-deep-adaptation/.

Nieuwenhout, Ceciel. 2021. 'Dutch Elections: Energy and Climate Considerations'. Personal Blog. *Energy and Climate Law*, 1 March. http://energyandclimatelaw .blogspot.com/2021/03/dutch-elections-energy-and -climate.html.

Nolan, Conor. 2021. '"Buy America": Protectionist Peril or Politically Practical?' *Georgetown Public Policy Review*, 31 October. https://gppreview.com/2021/10/31/buy -america-protectionist-peril-politically-practical/.

Nordhaus, William. 2015. 'Climate Clubs: Overcoming Free-Riding in International Climate Policy'. *American Economic Review* 105 (4): 1339–70. https://doi.org/10 .1257/aer.15000001.

Nye, Joseph S. 2004. *Soft Power: The Means to Success in World Politics*. New York: Public Affairs.

Oberthür, Sebastian. 2011. 'Global Climate Governance after Cancun: Options for EU Leadership'. *The International Spectator* 46 (1): 5–13. https://doi.org/10.1080/03932729 .2011.567900.

Oberthür, Sebastian, and Claire Dupont. 2021. 'The

European Union's International Climate Leadership: Towards a Grand Climate Strategy?' *Journal of European Public Policy* 28 (7): 1095–114. https://doi.org/10.1080 /13501763.2021.1918218.

Oberthür, Sebastian, and Claire Roche Kelly. 2008. 'EU Leadership in International Climate Policy: Achievements and Challenges'. *The International Spectator* 43 (3): 35–50. https://doi.org/10.1080/03932720802280594.

OECD. 2021. *Climate Finance Provided and Mobilised by Developed Countries: Aggregate Trends Updated with 2019 Data.* Climate Finance and the USD 100 Billion Goal. OECD. https://doi.org/10.1787/03590fb7-en.

Okyay, Asli Selin. 2019. 'EU–Turkey Migration Cooperation: From Saving the Day to Sustainable Mixed Migration Governance'. Working Paper 21. Global Turkey in Europe. Istituto Affari Internazionali. https://www.iai.it /sites/default/files/gte_wp_21.pdf.

O'Neil, Shannon K. 2021. 'America's Supply Chains Are Foreign Policy Now'. *Foreign Policy*, 16 February. https://foreignpolicy.com/2021/02/16/biden-supply-chains -manufacturing-foreign-policy-allies-globalization/.

Oomes, Nienke, and Katerina Kalcheva. 2007. 'Diagnosing Dutch Disease: Does Russia Have the Symptoms?' IMF Working Paper WP/07/102. International Monetary Fund. https://www.imf.org/external/pubs/ft/wp/2007/wp07102 .pdf.

O'Sullivan, Meghan, Indra Overland, and David Sandalow. 2017. 'The Geopolitics of Renewable Energy'. Working Paper. Columbia Center on Global Energy Policy & Harvard Belfer Center for Science and International Affairs. https://energypolicy.columbia.edu/sites/default /files/CGEPTheGeopoliticsOfRenewables.pdf.

Overland, Indra. 2019. 'The Geopolitics of Renewable Energy: Debunking Four Emerging Myths'. *Energy Research & Social Science* 49: 36–40. https://doi.org/10 .1016/j.erss.2018.10.018.

Parry, Ian, Simon Black, and James Roaf. 2021. 'Proposal for an International Carbon Price Floor among Large Emitters'. IMF Staff Climate Notes 2021/001. International

Monetary Fund. https://www.imf.org/en/Publications /staff-climate-notes/Issues/2021/06/15/Proposal-for-an -International-Carbon-Price-Floor-Among-Large-Emitters -460468.

Pastukhova, Maria, Jacopo Pepe, Kirsten Westphal, and Stiftung Wissenschaft und Politik. 2020. 'Beyond the Green Deal: Upgrading the EU's Energy Diplomacy for a New Era'. *SWP Comment*. https://doi.org/10.18449 /2020C31.

Piketty, Thomas. 2014. *Capital in the Twenty-First Century*. Cambridge, MA: The Belknap Press of Harvard University Press.

Pirani, Simon. 2021. 'Ukraine's Energy Policy and Prospects for the Gas Sector'. Energy Insight 106. Oxford Institute for Energy Studies. https://www.oxfordenergy .org/wpcms/wp-content/uploads/2021/12/Insight-106 -Ukraines-energy-policy-and-prospects-for-the-gas-sector .pdf.

Pistelli, Lapo. 2020. 'Addressing Africa's Energy Dilemma'. In: Manfred Hafner and Simone Tagliapietra (eds) *The Geopolitics of the Global Energy Transition* (Lecture Notes in Energy 73). Cham: Springer International Publishing, pp. 151–74.

Proedrou, Filippos. 2020. 'Anthropocene Geopolitics and Foreign Policy: Exploring the Link in the EU Case'. *Alternatives: Global, Local, Political* 45 (2): 83–101. https://doi.org/10.1177/0304375420931706.

Prokip, Andrian. 2019. 'Liberalizing Ukraine's Electricity Market: Benefits and Risks'. *Wilson Center – Kennan Institute* (blog), 6 May. https://www.wilsoncenter.org /blog-post/liberalizing-ukraines-electricity-market-benefits -and-risks.

Public Senat. 2019. 'La Crise des Gilets Jaunes en Dix Dates'. *Public Senat*, 30 March. https://www.publicsenat.fr /article/politique/la-crise-des-gilets-jaunes-en-dix-dates -139738.

Puig Cepero, Oriol, Sophie Desmidt, Adrien Detges, and Fabien Tondel. 2021. 'Climate Change, Development and Security in the Central Sahel'. Cascades. https://

www.cascades.eu/wp-content/uploads/2021/06/Climate
-Change-Development-and-Security-in-the-Central-Sahel
.pdf.

Putin, Vladimir. 2007. 'Speech and the Following Discussion
at the Munich Conference on Security Policy'. Presidency
of the Russian Federation. http://en.kremlin.ru/events
/president/transcripts/24034.

Rachman, Gideon. 2021. 'China Is Still a Long Way from Being
a Superpower'. *The Financial Times*, 19 July. https://www.ft
.com/content/bdaad457-9e22-4d74-b614-6cc44a613a0c.

Raimondi, Pier Paolo, and Simone Tagliapietra. 2021. 'The
Geopolitical Implications of Global Decarbonization
for MENA Producing Countries'. In OIES (ed.) *The
Geopolitics of Energy: Out with the Old, In with the
New?* OIES Forum, Issue 126. Oxford Institute for Energy
Studies. https://www.oxfordenergy.org/wpcms/wp-content
/uploads/2021/02/OEF-126.pdf.

Raineri, Luca. 2020. 'Sahel Climate Conflicts? When
(Fighting) Climate Change Fuels Terrorism'. Brief 20 –
Conflict Series. EU Institute for Security Studies. https://
www.iss.europa.eu/sites/default/files/EUISSFiles/Brief
%2020%20%20Sahel.pdf.

Raineri, Luca, and Alessandro Rossi. 2017. 'The Security–
Migration–Development Nexus in the Sahel: A Reality
Check'. IAI Working Papers 17 | 26. Istituto Affari
Internazionali.   https://www.iai.it/sites/default/files
/iaiwp1726.pdf.

Ren21. 2021. 'Renewables 2021: Global Status Report'.
Renewable Energy Policy Network for the 21st Century.
https://www.ren21.net/wp-content/uploads/2019/05
/GSR2021_Full_Report.pdf.

Reuters. 2021a. 'US Climate Envoy Kerry Urges China
to Keep Politics Out of Global Warming'. *Asia Pacific*,
9 February. https://www.reuters.com/world/asia-pacific
/china-holds-virtual-climate-meeting-with-us-describes
-environment-policy-oasis-2021-09-02/.

Reznik, Irina, and Henry Meyer. 2021. 'Putin Sees European
Gas Crisis as Russia's Golden Chance'. *Bloomberg*, 13
October. https://www.bloomberg.com/news/articles/2021

-10-13/putin-sees-european-gas-crisis-as-russia-s-golden
-opportunity.

Rifkind, Sir Malcolm. 2011. 'Europe Grapples with US
Pivot'. *The Diplomat*, 1 December. https://thediplomat
.com/2011/12/europe-grapples-with-u-s-pivot/.

Rosen, Amanda M. 2015. 'The Wrong Solution at the Right
Time: The Failure of the Kyoto Protocol on Climate
Change'. *Politics & Policy* 43 (1): 30–58. https://doi.org
/10.1111/polp.12105.

Rowlatt, Justin. 2021. 'COP26: World at One Minute
to Midnight over Climate Change – Boris Johnson'.
BBC News, 11 January. https://www.bbc.com/news/uk
-59114871.

Sabadus, Aura. 2021. 'Ukraine Can Play Key Role in Europe's
Energy Green Deal'. *Atlantic Council*, 20 February. https://
www.atlanticcouncil.org/blogs/ukrainealert/ukraine-can
-play-key-role-in-europes-energy-green-deal/.

Sandbu, Martin. 2021. 'The Imperceptible Approach of
Global Carbon Pricing'. *The Financial Times*, 11
November.   https://www.ft.com/content/ec9baebe-f81e
-4090-b1ec-909c9351a2b8.

Sapir, André. 2021. 'The European Union's Carbon Border
Mechanism and the WTO'. *Bruegel Blog*, 19 August.
https://www.bruegel.org/2021/07/the-european-unions
-carbon-border-mechanism-and-the-wto/.

Sapir, André, and Henrik Horn. 2020. 'Political Assessment
of Possible Reactions of EU Main Trading Partners to EU
Border Carbon Measures'. EP/EXPO/INTA/FWC/2019-01/
Lot5/1/C/02. European Parliament. https://data.europa.eu
/doi/10.2861/93094.

Schaller, Stella, and Alexander Carius. 2019. 'Convenient
Truths: Mapping Climate Agendas of Right-Wing
Populist Parties in Europe'. Adelphi Study. https://www
.adelphi.de/en/system/files/mediathek/bilder/Convenient
%20Truths%20-%20Mapping%20climate%20agendas
%20of%20right-wing%20populist%20parties%20in
%20Europe%20-%20adelphi.pdf.

Schmidt, Vivien A. 2013. 'Democracy and Legitimacy
in the European Union Revisited: Input, Output and

"Throughput"'. *Political Studies* 61 (1): 2–22. https://doi
.org/10.1111/j.1467-9248.2012.00962.x.

Schmitt, Thomas M. 2018. '(Why) Did Desertec Fail? An
Interim Analysis of a Large-Scale Renewable Energy
Infrastructure Project from a Social Studies of Technology
Perspective'. *Local Environment* 23 (7): 747–76. https://
doi.org/10.1080/13549839.2018.1469119.

Scholten, Daniel (ed.) 2018. *The Geopolitics of Renewables*
(Lecture Notes in Energy 61). Cham: Springer International
Publishing.

Serhan, Yasmeen. 2021. 'The Far-Right View on Climate
Politics'. *The Atlantic*, 10 August. https://www.theatlantic
.com/international/archive/2021/08/far-right-view-climate
-ipcc/619709/.

Shouse, Kate. 2021. 'US Climate Change Policy'. R46947.
CRS Report. US Congressional Research Service. https://
crsreports.congress.gov/product/pdf/R/R46947.

Simon, Frédéric. 2021. 'Leaked: The EU's Carbon Market
Reform Proposal'. *Euractiv*, 1 July. https://www.euractiv
.com/section/emissions-trading-scheme/news/leaked-the
-eus-carbon-market-reform-proposal/.

Sims Gallagher, Kelly. 2021. 'The Net Zero Trap: Countries
Need to Reduce Emissions Now, Not Just in the Distant
Future'. *Foreign Affairs*, 30 September. https://www
.foreignaffairs.com/articles/2021-09-30/net-zero-trap.

Sinn, Hans-Werner. 2012. *The Green Paradox: A Supply-Side
Approach to Global Warming*. Cambridge, MA: MIT
Press.

Spilimbergo, Antonio. 2021. 'Populism and Covid-19'.
VoxEU Columns, 13 August. https://voxeu.org/article
/populism-and-covid-19.

Steele, Paul. n.d. 'Why Adaptation Is the Greatest Market
Failure and What This Means for the State'. World
Resources Report. World Resources Institute. https://
www.wri.org/our-work/project/world-resources-report
/why-adaptation-greatest-market-failure-and-what-means
-state.

Stijn, Gabriel, and Barbara Rijks. 2020. 'Migration
Trends from, to and within the Niger (2016–2019)'.

PUB2020/104/R. UN Migration (Niger Data and Research Unit). https://publications.iom.int/system/files/pdf/iom-niger-four-year-report.pdf.

Stokes, Bruce. 2021. 'EU's Next Generation: It's the Climate, Stupid!' *Politico*, 9 October. https://www.politico.eu/article/eu-europe-next-generation-political-leaders-climate-change-emergency-afghanistan/.

Strambo, Claudia, Måns Nilsson, and André Månsson. 2015. 'Coherent or Inconsistent? Assessing Energy Security and Climate Policy Interaction within the European Union'. *Energy Research & Social Science* 8 (July): 1–12. https://doi.org/10.1016/j.erss.2015.04.004.

Stratfor. 2018. 'How Renewable Energy Will Change Geopolitics'. *RANE Worldview*, 27 June. https://worldview.stratfor.com/article/how-renewable-energy-will-change-geopolitics.

Szczepański, Marcin. 2020. 'Critical Raw Materials for the EU: Enablers of the Green and Digital Recovery'. Briefing. European Parliamentary Research Service. https://www.europarl.europa.eu/RegData/etudes/BRIE/2020/659426/EPRS_BRI(2020)659426_EN.pdf.

Tagliapietra, Simone, and Guntram Wolff. 2021. 'Relaunching Transatlantic Cooperation with a Carbon Border Adjustment Mechanism'. *Bruegel*, 6 November. https://www.bruegel.org/2021/06/relaunching-transatlantic-cooperation-with-a-carbon-border-adjustment-mechanism/.

Tagliapietra, Simone, and Georg Zachmann. 2021. 'Is Europe's Gas and Electricity Price Surge a One-Off?' *Bruegel Blog*, 13 September. https://www.bruegel.org/2021/09/is-europes-gas-and-electricity-price-surge-a-one-off/.

Tagliapietra, Simone, Georg Zachmann, Ottmar Edenhofer, and Jean-Michel Glachant. 2019. 'The European Union Energy Transition: Key Priorities for the Next Five Years'. Policy Brief, Issue 1. https://www.bruegel.org/wp-content/uploads/2019/07/Bruegel_Policy_Brief-2019_01.pdf.

Tanchum, Michaël, and Theodore Murphy. 2021. 'The EU's Global Gateway and a New Foundation for Partnerships in Africa'. *European Council on Foreign Relations*, 29

September. https://ecfr.eu/article/the-eus-global-gateway -and-a-new-foundation-for-partnerships-in-africa/.

Tänzler, Dennis, Sebastian Oberthür, and Emily Wright. 2020. 'The Geopolitics of Decarbonisation: Reshaping European Foreign Relations'. Institute for European Studies & Adelphi. https://climate-diplomacy.org/magazine /cooperation/geopolitics-decarbonisation-reshaping -european-foreign-relations.

Teevan, Chloe, Alfonso Medinilla, and Katja Sergejeff. 2021. 'The Green Deal in EU Foreign and Development Policy'. Briefing Note 131. European Centre for Development Policy Management. https://ecdpm.org/wp-content /uploads/Green-Deal-EU-Foreign-Development-Policy -ECDPM-Briefing-Note-131-2021.pdf.

The Economist. 2014. 'Europe's Ring of Fire'. The Economist, 20 September. https://www.economist.com/europe/2014 /09/20/europes-ring-of-fire.

The Economist. 2017. 'The Importance of a European Foreign and Security Policy'. The Economist, 23 March. https://www.economist.com/special-report/2017/03/23 /the-importance-of-a-european-foreign-and-security -policy.

The Economist. 2021. 'The Climate Has Overtaken Covid-19 as German Voters' Top Concern'. The Economist, 24 September, sec. Daily Chart. https:// www.economist.com/graphic-detail/2021/09/24/the-climate- has-overtaken-covid-19-as-german-voters-top-concern.

Thebault, Reis. 2021. '"Winter Is Coming": EU Urges Members to Protect the Poor while Tackling High Energy Costs'. The Washington Post, 13 October. https://www .washingtonpost.com/world/europe/eu-energy-crisis/2021 /10/13/918f6154-2b97-11ec-b17d-985c186de338_story .html.

Timperley, Jocelyn. 2021. 'The Broken $100-Billion Promise of Climate Finance — and How to Fix It'. Nature 598 (7881): 400–402. https://doi.org/10.1038/d41586-021 -02846-3.

Tiounine, Margot, and Tom Hannen. 2019. 'Vladimir Putin: The Full Interview'. The Financial Times, 5 August.

https://www.ft.com/video/d62ed062-0d6a-4818-86ff
-4b8120125583.

Tocci, Nathalie. 2007. *The EU and Conflict Resolution:
Promoting Peace in the Backyard* (Routledge UACES
Contemporary European Studies 1). London: Routledge.

Tocci, Nathalie. 2019a. 'Navigating Complexity: The EU's
Rationale in the 21st Century'. *IAI Commentaries* 19
(6): 1–5. https://www.iai.it/en/pubblicazioni/navigating
-complexity-eus-rationale-21st-century.

Tocci, Nathalie. 2019b. 'Academia and Practice in European
Foreign Policy: Opportunities for Mutual Learning'.
Journal of European Integration 40 (7): 837–52. https://
doi.org/10.1080/07036337.2018.1524466.

Tocci, Nathalie. 2019c. 'Resilience and the Role of the
European Union in the World'. *Contemporary Security
Policy* 41 (2): 176–94. https://doi.org/10.1080/13523260
.2019.1640342.

Tocci, Nathalie. 2020. 'International Order and the European
Project in Times of COVID19'. *IAI Commentaries* 20 (9):
1–7. https://www.iai.it/sites/default/files/iaicom2009.pdf.

Tocci, Nathalie. 2021a. 'European Strategic Autonomy:
What It Is, Why We Need It, How to Achieve It'. Istituto
Affari Internazionali. https://www.iai.it/sites/default/files
/9788893681780.pdf.

Tocci, Nathalie. 2021b. 'Teetering on the Brink: Turkey's
Troubled Ties with the West'. *Journal of Middle
Eastern Politics and Policy* 2021: 20–26. https://jmepp
.hkspublications.org/2021-fall-edition/.

Tocci, Nathalie. 2021c. 'Europe's Anxieties about Biden
Are Really Anxieties about Itself'. *Politico*, 6 September.
https://www.politico.eu/article/europe-joe-biden-anxiety/.

Tocci, Nathalie and Riccardo Alcaro. 2014. 'Rethinking
Transatlantic Relations in a Multipolar Era'. *International
Politics* 51: 366–89.

Tocci, Nathalie, and Senem Aydın-Düzgit. 2015. *Turkey
and the European Union*. New York, NY: Palgrave
Macmillan.

Tocci, Nathalie, Hakim Darbouche, Michael Emerson,
and Sandra Fernandes. 2008. 'The European Union

as a Normative Foreign Policy Actor'. CEPS Working Document 281. https://www.ceps.eu/ceps-publications /european-union-normative-foreign-policy-actor/.

Tooze, Adam. 2021. 'Present at the Creation of a Climate Alliance – or Climate Conflict'. *Foreign Policy*, 6 August. https://foreignpolicy.com/2021/08/06/climate-conflict -europe-us-green-trade-war/.

Tran, Hung. 2021. 'Decoupling/Reshoring versus Dual Circulation: Competing Strategies for Security and Influence'. Issue Brief. Atlantic Council – Geoeconomics Center.   https://www.atlanticcouncil.org/wp-content /uploads/2021/04/Decoupling_Reshoring_versus_Dual _Circulation.pdf.

Treib, Oliver. 2021. 'Euroscepticism Is Here to Stay: What Cleavage Theory Can Teach Us about the 2019 European Parliament Elections'. *Journal of European Public Policy* 28 (2): 174–89. https://doi.org/10.1080/13501763.2020 .1737881.

Ulgen, Sinan, Mehveş Selamoğlu, and Azem Yıldırım. 2021. 'Modernizing the EU–Turkey Customs Union: The Digital Agenda and the Green Deal'. Carnegie Europe.

UNDP. 2021. 'Nationally Determined Contributions (NDC) Global Outlook Report 2021: The State of Climate Ambition'. United Nations Development Programme. https://www.undp.org/sites/g/files/zskgke326/files/2021 -11/UNDP-NDC-Global-Outlook-Report-2021-The-State -of-Climate-Ambition-V2.pdf.

UNEP. 2021. 'Adaptation Gap Report 2020'. UN Environment Programme. https://wedocs.unep.org/bitstream/handle/20 .500.11822/34721/AGR2020.pdf.

US-DOS. 2021a. 'Joint Statement of the United States and Germany on Support for Ukraine, European Energy Security, and Our Climate Goals'. US Department of State – German Foreign Ministry. https://www.state.gov/joint -statement-of-the-united-states-and-germany-on-support -for-ukraine-european-energy-security-and-our-climate -goals/.

US-DOS. 2021b. 'US–China Joint Statement Addressing the Climate Crisis'. US Department of State – Chinese

Government. https://www.state.gov/u-s-china-joint -statement-addressing-the-climate-crisis/.

US-DOS. 2021c. 'US–China Joint Glasgow Declaration on Enhancing Climate Action in the 2020s'. US Department of State – Chinese Government. https:// www.state.gov/u-s-china-joint-glasgow-declaration-on-enhancing-climate-action-in-the-2020s/.

Vandenbroucke, Frank, Catherine Barnard, and Geert de Baere (eds) 2017. *A European Social Union after the Crisis*. Cambridge: Cambridge University Press.

Vanheukelen. 2021. 'EU Climate Diplomacy: Projecting Green Global Leadership'. EU Diplomacy Paper. College of Europe. https://www.coleurope.eu/sites/default/files /uploads/news/EDP%206%202021_Vanheukelen.pdf.

Venturi, Bernardo. 2017. 'The EU and the Sahel: A Laboratory of Experimentation for the Security–Migration–Development Nexus'. IAI Working Papers 17 | 38. Istituto Affari Internazionali. https://www.iai.it/sites/default/files /iaiwp1738.pdf.

Venturi, Bernardo, and Luca Barana. 2021. 'Lake Chad: Another Protracted Crisis in the Sahel or a Regional Exception?' IAI Papers 21 | 10. Istituto Affari Internazionali. https://www.iai.it/sites/default/files/iaip2110.pdf.

Venturi, Riccardo, Lorenzo Colantoni, and Daniele Fattibene. 2020. 'GhanAgri'. Istituto Affari Internazionali. https:// www.iai.it/sites/default/files/ghanagri.pdf.

Washington Post. 2021. 'The Glasgow Climate Pact, Annotated'. 13 November. https://www.washingtonpost .com/climate-environment/interactive/2021/glasgow -climate-pact-full-text-cop26/.

Weko, Silvia, Laima Eicke, Adela Marian, and Maria Apergi. 2020. 'The Global Impacts of an EU Carbon Border Adjustment Mechanism'. Institute for Advanced Sustainability Studies (IASS). https://publications.iass -potsdam.de/pubman/item/item_6000630.

Werz, Michael, and Laura Conley. 2012. 'Climate Change, Migration, and Conflict in Northwest Africa: Rising Dangers and Policy Options across the Arc of Tension'. Tackling Climate Change and Environmental Injustice.

American Progress (CAP). https://cdn.americanprogress
.org/wp-content/uploads/issues/2012/04/pdf/climate
_migration_nwafrica.pdf.

Werz, Michael, and Max Hoffman. 2016. 'Europe's Twenty-
First Century Challenge: Climate Change, Migration and
Security'. *European View* 15 (1): 145–54. https://doi.org
/10.1007/s12290-016-0385-7.

White, Sarah, and Victor Mallet. 2021. 'France: The Battle over
Wind Power Stirs Up the Election'. *The Financial Times*,
2 December. https://www.ft.com/content/29cb5f2b-9b09
-49bf-b306-c3a782191f6c.

WindEurope. 2020. 'Financing and Investment Trends
2020'. *WindEurope*, 13 April. https://windeurope.org
/intelligence-platform/product/financing-and-investment
-trends-2020/.

Wondreys, Jakub, and Cas Mudde. 2020. 'Victims of the
Pandemic? European Far-Right Parties and COVID-19'.
Nationalities Papers 50 (1): 86–103. https://doi.org/10
.1017/nps.2020.93.

World Bank. 2021. 'Global Gas Flaring Tracker Data:
Individual Flare Sites – Gas Flaring Volumes (m/n m^3/yr)
for 2020'. The World Bank Group. https://www.ggfrdata
.org.

WWF Europe et al. 2021. 'Open Letter to the Commission:
Asking for Removal of Forestry and Bioenergy
from the Current Climate Taxonomy Delegated
Act'. WWF European Policy Office & alii. https://
wwfeu.awsassets.panda.org/downloads/open_letter_to_european
_commission_forestry_bioenergy_and_taxonomy_april21
.pdf.

Xu, Chi, Timothy A. Kohler, Timothy M. Lenton,
Jens-Christian Svenning, and Marten Scheffer. 2020.
'Future of the Human Climate Niche'. *Proceedings of
the National Academy of Sciences* 117 (21): 11350–55.
https://doi.org/10.1073/pnas.1910114117.

Yergin, Daniel. 2021. 'Why the Energy Transition Will Be So
Complicated'. *The Atlantic*, 27 November. https://www
.theatlantic.com/international/archive/2021/11/energy
-shock-transition/620813/.

Youngs, Richard. 2015. *Climate Change and European Security*. New York: Routledge.

Youngs, Richard. 2020. 'EU Foreign Policy and Energy Strategy: Bounded Contestation'. *Journal of European Integration* 42 (1): 147–62. https://doi.org/10.1080/07036337.2019.1708345.

Zachmann, Georg, Gustav Fredrikson, and Grégory Claeys. 2018. 'The Distributional Effects of Climate Policies'. Blueprint Series 28. Bruegel. https://www.bruegel.org/wp-content/uploads/2018/11/Bruegel_Blueprint_28_final1.pdf.

Żuk, Piotr, and Kacper Szulecki. 2020. 'Unpacking the Right-Populist Threat to Climate Action: Poland's Pro-Governmental Media on Energy Transition and Climate Change'. *Energy Research & Social Science* 66: 101485. https://doi.org/10.1016/j.erss.2020.101485.

Index